Compendium of Terminology and Nomenclature of Properties in
Clinical Laboratory Sciences
Recommendations 2016

International Union of Pure and Applied Chemistry
Chemistry and Human Health Division (VII)
Subcommittee on Nomenclature for Properties and Units

International Federation of Clinical Chemistry and Laboratory Medicine
Scientific Division (IX)
Committee on Nomenclature for Properties and Units

Compendium of Terminology and Nomenclature of Properties in Clinical Laboratory Sciences
Recommendations 2016

Georges Férard
University of Strasbourg, France
Email: georges.ferard@noos.fr

René Dybkaer
Frederiksberg Hospital, Denmark
Email: rene.dybkaer@frh.regionh.dk

and

Xavier Fuentes-Arderiu
Clinical Laboratory Sciences C, Barcelona, Spain
Email: 2461xfa@gmail.com

ROYAL SOCIETY OF CHEMISTRY

iFCC
International Federation
of Clinical Chemistry
and Laboratory Medicine

THE QUEEN'S AWARDS
FOR ENTERPRISE:
INTERNATIONAL TRADE
2013

Print ISBN: 978-1-78262-107-2
PDF eISBN: 978-1-78262-245-1
EPUB eISBN: 978-1-78262-909-2

A catalogue record for this book is available from the British Library

Published by The Royal Society of Chemistry,
Thomas Graham House, Science Park, Milton Road,
Cambridge CB4 0WF, UK

Registered Charity Number 207890

Visit our website at www.rsc.org/books

Printed in the United Kingdom by CPI Group (UK) Ltd, Croydon, CR0 4YY, UK

Preface

Over the last 60 years, much effort has been made to introduce and apply in the clinical laboratory sciences the concepts, designations, rules, and conventions on properties, including quantities and units, recommended by international organizations such as CGPM, ISO, IUPAC and IFCC.

From 1994, extensions and applications to several disciplines within the clinical laboratory sciences have been made, by the IFCC/IUPAC Committee/Subcommittee on Nomenclature for Properties and Units (C-SC-NPU).

In 1995, the first issue of the Silver Book was published to harmonize and facilitate access to relevant documents. From this time, many recommendations and technical reports have been prepared by the C-SC-NPU, but they are not readily available and some have been updated and aligned with other documents.

IUPAC and IFCC have now decided that, after 20 years, it is time to issue a second edition of the Silver Book with four objectives:

- to update the recommendations and technical reports;
- to enlarge the subject field by several disciplines applied in the clinical laboratory sciences;
- to develop concepts used to include properties having no quantity dimensions, that are frequently submitted to examination in clinical laboratories; and
- to explain when necessary the recommendations and illustrate them by examples taken from laboratory practice.

Comments and suggestions are welcomed.

The authors wish to thank members of the Committee/Subcommittee on Nomenclature for Properties and Units (IFCC and IUPAC): Ivan Bruunshuus (Alleroed); Pedro de Araujo (Sao Paulo); Robert Flatman (Taringa, Chair); Urban Forsum (Linköping); Gilbert Hill (Toronto); Antonin Jabor (Kladno); Jens Gledisch (Oslo); Helle Johannessen (Copenhagen); Daniel Karlsson (Linköping); Wolf Külpmann (Hannover); Ulla Magdal-Petersen (Copenhagen); Clement McDonald (Indianapolis); Gunnar Nordin (Uppsala); Henrik Olesen (Copenhagen); Françoise Pontet (Paris); Gunther Schadow (Indianapolis); and Kaoru Yamauchi (Tokyo).

Georges Férard, René Dybkaer and Xavier Fuentes-Arderiu

Comments to:
Georges Férard: georges.ferard@noos.fr

Compendium of Terminology and Nomenclature of Properties in Clinical Laboratory Sciences: Recommendations 2016
Edited by Georges Férard, René Dybkaer and Xavier Fuentes-Arderiu
© International Union of Pure and Applied Chemistry 2017
Published by the Royal Society of Chemistry, www.rsc.org

Committees

This second edition of the Compendium of Terminology and Nomenclature of Properties in Clinical Laboratory Sciences was discussed in an IFCC and IUPAC joint Working Group: René Dybkaer (Frederiksberg); Georges Férard (Strasbourg, Chair); Françoise Pontet (Paris, co-Chair); Xavier Fuentes-Arderiu (Barcelona); Dongchon Kang (Kyushu); Gilbert Hill (Toronto); Clement McDonald (Indianapolis); and Anders J. Thor (Stockholm).

The Working Group wishes to thank members of the C-SC-NPU (IFCC and IUPAC): Ivan Bruunshuus (Alleroed); Pedro de Araujo (Sao Paulo); Robert Flatman (Taringa, Chair); Urban Forsum (Linköping); Gilbert Hill (Toronto); Antonin Jabor (Kladno); Jens Gledisch (Oslo); Helle Johannessen (Copenhagen); Daniel Karlsson (Linköping); Wolf Külpmann (Hannover); Ulla Magdal-Petersen (Copenhagen); Clement McDonald (Indianapolis); Gunnar Nordin (Uppsala); Henrik Olesen (Copenhagen); Françoise Pontet (Paris); Gunther Schadow (Indianapolis); and Kaoru Yamanouchi (Tokyo).

The first edition of the Compendium (1995) was authored by J. Christopher Rigg, Stanley S. Brown, René Dybkaer, and Henrik Olesen.

Memberships of the Committees

International Union of Pure and Applied Chemistry

The membership of the Commission on Quantities and Units, subsequently Subcommittee on Nomenclature for Properties and Units, during the period 1968 to 2013, when the successive recommendations and technical reports on properties, quantities and units were prepared was as follows:

Chairholders: 1968–1975 R. Dybkær (Denmark); 1976–1979 R. Zender (Switzerland); 1980–1989 H. P. Lehmann (United States); 1989–1995 H. Olesen (Denmark); 1996–1997 D. Kenny (Ireland); 1998–2000 X. Fuentes-Arderiu (Spain); 2001–2005 H. Forsum (Sweden); 2006–2011 F. Pontet (France); and 2012– R. Flatman (Australia).

Titular members and consultants: 1968–1975 B. H. Armbrecht (United States); 1983–1991 D. R. Bangham (United Kingdom); 1983–1987 L. F. Bertello (Argentina); 2004– I. Bruunshuus Petersen (Denmark); 1968–1977 and 2002– R. Dybkær (Denmark); 1983–1991 and 2008– G. Férard (France); 2010– R. Flatman (Australia); 1998– U. Forsum (Sweden); 1992–1995 and 2007– X. Fuentes-Arderiu (Spain); 1971–1979 R. Herrmann (Germany); 1987–1995 and 2010– J. G. Hill (Canada); 2005–2007 J. Ihalainen (Finland); 1998–2007 A. Jabor (Czech Republic); 1968–1973 K. Jørgensen (Denmark); 2006–2008 D. Karlsson (Sweden); 2008– D. Kang (Japan); 1998–2007 D. Kenny (Ireland); 1985–1993 M. Lauritzen (Denmark); 1979–1989 H. P. Lehmann (United States); 2008– C. McDonald (United States); 1968–1975 P. Métais

Compendium of Terminology and Nomenclature of Properties in Clinical Laboratory Sciences:
Recommendations 2016
Edited by Georges Férard, René Dybkaer and Xavier Fuentes-Arderiu
© International Union of Pure and Applied Chemistry 2017
Published by the Royal Society of Chemistry, www.rsc.org

(France); 2000– G. Nordin (Sweden); 1988–1995 and 2006–2007 H. Olesen (Denmark); 1975–1979 C. Onkelinx (Belgium); 2006– U. M. Petersen (Denmark); 1973–1977 and 1986–1989 J. C. Rigg (the Netherlands); 2008– G. Schadow (United States); 1994–2001 P. Soares de Araujo (Brazil); 2002–2007 H. Storm (the Netherlands); 1977–1981 B. F. Visser (the Netherlands); and 1975–1979 R. Zender (Switzerland).

International Federation of Clinical Chemistry and Laboratory Medicine

The membership of the Committee on Quantities and Units, subsequently Committee on Nomenclature for Properties and Units during the period 1968 to 2013, when the successive recommendations and technical on properties, quantities and units were prepared was as follows:

Chairholders: 1968–1975 R. Dybkær (Denmark); 1976–1979 R. Zender (Switzerland); 1980–1989 H. P. Lehmann (United States); 1989–1995 H. Olesen (Denmark); 1996–1997 D. Kenny (Ireland); 1998–2000 X. Fuentes-Arderiu (Spain); 2001–2005 U. Forsum (Sweden); 2006–2011 F. Pontet (France); and 2012– R. Flatman (Australia).

Titular members and consultants: 1968–1975 B. H. Armbrecht (United States); 1983–1991 D. R. Bangham (United Kingdom); 1983–1987 L. F. Bertello (Argentina); 2005– I. Bruunshuus Petersen (Denmark); 1968–1977 and 1996–1998 and 2000– R. Dybkær (Denmark); 1983–1991 G. Férard (France); 2010– R. Flatman (Australia); 2006– U. Forsum (Sweden); 1992–1995 and 2006– X. Fuentes-Arderiu (Spain); 1971–1979 R. Herrmann (Germany); 1988–1999 J. G. Hill (Canada); 2005–2007 J. Ihalainen (Finland); 1999–2004 A. Jabor (Czech Republic); 1968–1973 K. Jørgensen (Denmark); 2006–2008 D. Karlsson (Sweden); 1985–1993 M. Lauritzen (Denmark); 1979–1989 H. P. Lehmann (United States); 1996–1999 C. McDonald (United States); 1968–1975 P. Métais (France); 2000–2005 G. Nordin (Sweden); 1988–1995 H. Olesen (Denmark); 1975–1979 C. Onkelinx (Belgium); 2006– U. M. Petersen (Denmark); 1973–1977 and 1986–1989 J. C. Rigg (the Netherlands); 1975–1979 O. Siggaard-Andersen (Denmark); 1996–2004 P. Soares de Araujo (Brazil); 1991–1994 P. Storring (United Kingdom); 1977–1981 B. F. Visser (the Netherlands); and 1975–1979 R. Zender (Switzerland).

The membership of Working Group on the Silver Book Revision during the period 2008 to 2015 was as follows:

Chairholders: G. Férard (France) and F. Pontet (†) (France)

Members and consultants: R. Dybkaer (Denmark); X. Fuentes-Arderiu (Spain); G. Hill (Canada); D. Kang (Japan); C. McDonald (United States); and A. Thor (†) (Sweden).

List of Abbreviations Used for Institutions and Committees

BCR	Bureau Communautaire de Référence (of European Communities)
BIML	International Bureau of Legal Metrology/Bureau International de Métrologie Légale
BIPM	International Bureau of Weights and Measures/Bureau International des Poids et Mesures
CCQM	Consultative Committee for Amount of Substance – Metrology in Chemistry and Biology (formerly Consultative Committee for Amount of Substance)/Comité Consultatif pour la Quantité de Matière – Métrologie en Chimie et Biologie (of CIPM)
CCTF	Consultative Committee for Time and Frequency (of CIPM)
CCU	Consultative Committee for Units/Comité Consultatif d'Unités (of CIPM)
CEC	Commission of the European Communities
CEN	European Committee for Standardization/Comité Européen de Normalisation
CGPM	General Conference on Weights and Measures/Conférence Générale des Poids et Mesures
CIAAW	Commission on Isotopic Abundances and Atomic Weights (of IUPAC) (formerly CAWIA)
CIPM	International Committee for Weights and Measures/Comité International des Poids et Mesures
CITAC	Cooperation on International Traceability in Analytical Chemistry
CLSI	Clinical and Laboratory Standards Institute (formerly NCCLS)
CODATA	Committee on Data for Science and Technology
EA	European co-operation for Accreditation
ECBS	Expert Committee on Biological Standardization (of WHO)
ECCLS	European Committee for Clinical Laboratory Standards
EFCC	European Federation of Clinical Chemistry and Laboratory Medicine
EMBO	European Molecular Biology Organization
EURAMET	European Association of National Metrology Institutes (formerly EUROMET)
IAEA	International Atomic Energy Agency
ICRU	International Commission on Radiation Units and Measurements
ICSB	International Committee on Systematic Bacteriology
ICSH	International Council for Standardization in Haematology (formerly International Committee for Standardization in Haematology)

Compendium of Terminology and Nomenclature of Properties in Clinical Laboratory Sciences: Recommendations 2016
Edited by Georges Férard, René Dybkaer and Xavier Fuentes-Arderiu
© International Union of Pure and Applied Chemistry 2017
Published by the Royal Society of Chemistry, www.rsc.org

ICSU	International Council for Science (formerly International Council of Scientific Unions)
ICTNS	International Committee on Terminology, Nomenclature and Symbols (of IUPAC)
ICVGAN	International Committee on Veterinary Gross Anatomical Nomenclature
IDCNS	Interdivisional Committee on Nomenclature and Symbols (of IUPAC) (superseded by ICTNS)
IEC	International Electrotechnical Commission/Commission électrotechnique internationale
IFCC	International Federation of Clinical Chemistry and Laboratory Medicine
ILAC	International Laboratory Accreditation Cooperation
IOC	International Olympic Committee
IRMM	Institute for Reference Materials and Measurements/Institut des Matériaux et Mesures de Référence
ISA	International Society of Andrology
ISO	International Organization for Standardization
ISO/TC12	ISO Technical Committee on Quantities, Symbols, and Conversion Factors
ISTH	International Society on Thrombosis and Haematology
IUB	International Union of Biochemistry (superseded by IUBMB)
IUBMB	International Union of Biochemistry and Molecular Biology (formerly IUB)
IUIS	International Union of Immunological Societies
IUMS	International Union of Microbiological Societies
IUNS	International Union of Nutritional Sciences
IUPAC	International Union of Pure and Applied Chemistry
IUPAP	International Union of Pure and Applied Physics
IUPS	International Union of Physiological Sciences
JCGM	Joint Committee for Guides in Metrology
JCGM/WG1	Joint Committee for Guides in Metrology, Working Group 1 on the GUM (Guide to the Expression of Uncertainty in Measurement)
JCGM/WG2	Joint Committee for Guides in Metrology, Working Group 2 on the VIM (International vocabulary of metrology – Basic and general concepts and associated terms)
JCTLM	Joint Committee for Traceability in Laboratory Medicine
NCCLS	National Committee for Clinical Laboratory Standards (superseded by CLSI)
NIST	US National Institute of Science and Technology
OIML	International Organization of Legal Metrology/Organisation Internationale de Métrologie Légale
REMCO	Committee on Reference Materials (of ISO)
WASP	World Association of (Anatomic and Clinical) Pathological Societies
WHO	World Health Organization/Organisation mondiale de la Santé

List of Symbols, Terms and SI units for Kinds-of-quantity

A given symbol is sometimes used for different kinds-of-quantity. Different symbols are sometimes used for a given kind-of-quantity. Synonyms for a kind-of-quantity are included unless they are deprecated. Many kinds-of-quantity can be expressed in other coherent SI units than those listed below in column 3.

Symbol(s) of kind-of-quantity	Term(s) for kind(s)-of-quantity	Symbol(s) for SI base unit or coherent derived SI unit
a	relative chemical activity	1
a_b	relative molal activity	1
a_c	relative substance-concentrational activity	1
a_x	relative substance-fractional activity	1
$\boldsymbol{a}, a, \vec{a}$	acceleration, linear acceleration	$\mathrm{m\ s^{-2}}$
$\boldsymbol{a}_{rot}$	centrifugal acceleration	$\mathrm{m\ s^{-2}}$
a_l, a, K	lineic decadic absorbance, linear decadic absorption coefficient	$\mathrm{m^{-1}}$
a	massic area	$\mathrm{m^2\ kg^{-1}}$
a, D_T	thermal diffusivity, thermal diffusion coefficient	$\mathrm{m^2\ s^{-1}}$
a_1, f_1	massic Helmholtz energy, massic Helmholtz free energy, massic Helmholtz function	$\mathrm{J\ kg^{-1}} = \mathrm{m^2\ s^{-2}}$
a	absorbed dose content	$\mathrm{Bq\ kg^{-1}} = \mathrm{kg^{-1}\ s^{-1}}$
A	decadic absorbance	1
A, S	area	$\mathrm{m^2}$
A, F	Helmholtz energy, Helmholtz free energy, Helmholtz function	$\mathrm{J} = \mathrm{kg\ m^2\ s^{-2}}$
A	activity referred to a radionuclide, absorbed dose	$\mathrm{Bq} = \mathrm{s^{-1}}$
A	molar Gibbs energy, affinity	$\mathrm{J\ mol^{-1}} = \mathrm{kg\ m^2\ mol^{-1}\ s^{-2}}$
$A(\Delta\lambda)$	integral of molar area Napierian absorbance over wavelength	$\mathrm{m^3\ mol^{-1}}$
A_m	molar area	$\mathrm{m^2\ mol^{-1}}$
A_m	molar absorbed dose	$\mathrm{Bq\ mol^{-1}} = \mathrm{s^{-1}\ mol^{-1}}$

Compendium of Terminology and Nomenclature of Properties in Clinical Laboratory Sciences: Recommendations 2016
Edited by Georges Férard, René Dybkaer and Xavier Fuentes-Arderiu
© International Union of Pure and Applied Chemistry 2017
Published by the Royal Society of Chemistry, www.rsc.org

Symbol(s) of kind-of-quantity	Term(s) for kind(s)-of-quantity	Symbol(s) for SI base unit or coherent derived SI unit
b, y	breadth	m
b, m	molality	mol kg^{-1}
$\tilde{b}, \tilde{m}$	active molality	mol kg^{-1}
b_E, κ_E, e_E	catalytic-activity concentration, catalytic concentration	$\text{kat m}^{-3} = \text{mol m}^{-3}\ \text{s}^{-1}$
B	napierian absorbance of radiation	1
$\boldsymbol{B}$	magnetic induction, magnetic flux density	$\text{kg s}^{-2}\ \text{A}^{-1}$
$\boldsymbol{c}, \boldsymbol{v}, \boldsymbol{u}, \boldsymbol{w}$	length rate vector, velocity	m s^{-1}
c	length rate of electromagnetic radiation, speed of propagation of electromagnetic radiation	m s^{-1}
c	substance concentration, amount-of-substance concentration, amount concentration, molar concentration	mol m^{-3}
c	massic kelvic enthalpy, massic heat capacity	$\text{J kg}^{-1}\ \text{K}^{-1} = \text{m}^2\ \text{s}^{-2}\ \text{K}^{-1}$
$\tilde{c}$	active substance concentration	mol m^{-3}
$\hat{c}$	osmolarity, osmotic concentration	mol m^{-3}
C	number concentration	m^{-3}
C	clearance	$\text{m}^3\ \text{s}^{-1}$
C	capacitance, electrical capacitance, electric capacitance	$\text{F} = \text{kg}^{-1}\ \text{m}^{-2}\ \text{s}^4\ \text{A}^2$
C	kelvic enthalpy, heat capacity	$\text{J K}^{-1} = \text{kg m}^2\ \text{s}^{-2}\ \text{K}^{-1}$
C_m	molar kelvic enthalpy, molar heat capacity	$\text{J K}^{-1}\ \text{mol}^{-1} = \text{kg m}^2\ \text{s}^{-2}\ \text{K}^{-1}\ \text{mol}^{-1}$
d	relative volumic mass, relative mass density	1
d, r	distance travelled or migrated	m
d, D	diameter of circle	m
$d, \delta, -z$	depth	m
$D_i(\lambda), D(\lambda)$	attenuance, extinction factor	1
D, d	diameter of circle	m
D	massic energy of ionizing radiation absorbed, massic absorbed dose	$\text{Gy} = \text{m}^2\ \text{s}^{-2}$
D	diffusion coefficient	$\text{m}^2\ \text{s}^{-1}$
D^{-1}	resistance to diffusion	s m^{-2}
D_T, a	thermal diffusivity, thermal diffusion coefficient	$\text{m}^2\ \text{s}^{-1}$
$\dot{D}$	massic energy rate of ionizing radiation, massic absorbed dose	$\text{Gy s}^{-1} = \text{m}^2\ \text{s}^{-3}$
e	massic energy	$\text{J kg}^{-1} = \text{m}^2\ \text{s}^{-2}$
e	electrical charge constant, elementary charge	$\text{C} = \text{A s}$
e_E, b_E, κ_E	catalytic-activity concentration, catalytic concentration	$\text{kat m}^{-3} = \text{mol m}^{-3}\ \text{s}^{-1}$
E, E_{mf}	electromotive force	$\text{V} = \text{kg m}^2\ \text{s}^{-3}\ \text{A}^{-1}$
$E, \boldsymbol{E}$	electric field strength	$\text{V m}^{-1} = \text{kg m A}^{-1}\ \text{s}^{-3}$
E, Q	energy, amount-of-energy	$\text{J} = \text{kg m}^2\ \text{s}^{-2}$
E_e	areic energy rate of radiation, irradiance	$\text{W m}^{-2} = \text{kg s}^{-3}$
E_k, E_{kin}, T	kinetic energy	$\text{J} = \text{kg m}^2\ \text{s}^{-2}$
E_m	molar activation energy	$\text{J mol}^{-1} = \text{kg m}^2\ \text{mol}^{-1}\ \text{s}^{-2}$
E_{pot}, V, Φ	potential energy	$\text{J} = \text{kg m}^2\ \text{s}^{-2}$

Symbol(s) of kind-of-quantity	Term(s) for kind(s)-of-quantity	Symbol(s) for SI base unit or coherent derived SI unit
f	activity factor, activity coefficient	1
f	osmotic factor, osmotic coefficient	1
$f, v, \Phi_N, \dot{N}$	number rate	s^{-1}
f, a	massic Helmholtz energy, massic Helmholtz free energy, massic Helmholtz function	$J\,kg^{-1} = m^2\,s^{-2}$
$f, \tilde{p}$	active partial pressure, fugacity	$Pa = kg\,m^{-1}\,s^2$
f, v	number rate of regular event, frequency	$Hz = s^{-1}$
f, V	electric potential	$V = kg\,m^2\,s^{-3}\,A^{-1}$
f_{rot}	number rate of rotation, rotational frequency	$Hz = s^{-1}$
f_x, γ_x	substance-fractional activity factor, rational activity coefficient	1
$F_v, \dot{V}, \Phi_v$	volume rate, volume flow rate, volume rate of flow	$m^3\,s^{-1}$
F	Faraday constant	$C\,mol^{-1} = A\,s\,mol^{-1}$
F	heat flow rate	$W = kg\,m^2\,s^{-3}$
F, A	Helmholtz energy, Helmholtz free energy, Helmholtz function	$J = kg\,m^2\,s^{-2}$
$F_m, \dot{m}, \Phi_m, J_m$	mass rate, mean mass rate, mass flow rate, mass rate of flow, mass transfer rate, mass velocity	$kg\,s^{-1}$
$F, \mathbf{F}, \vec{F}$	force	$N = kg\,m\,s^{-2}$
$F_g, \mathbf{F}_g, \vec{F}_g$	force due to gravity	$N = kg\,m\,s^{-2}$
$F_n, \Phi_n, \dot{n}$	substance rate, substance flow rate	$mol\,s^{-1}$
$F_{rot}, \mathbf{F}_{rot}$	centrifugal force	$N = kg\,m\,s^{-2}$
F'	counterforce	$N = kg\,m\,s^{-2}$
g	activity factor	1
$g, \mathbf{g}, \vec{g}$	acceleration due to gravity, acceleration by gravity, acceleration of free fall	$m\,s^{-2}$
g	massic Gibbs energy, massic Gibbs free energy	$J\,kg^{-1} = m^2\,s^{-2}$
$\mathbf{grad}_x\,c, \nabla c$	substance concentration gradient	$mol\,m^{-4}$
$\mathbf{grad}\,C, \nabla C$	number concentration gradient	m^{-4}
$\mathbf{grad}\,p, \nabla p$	volumic mass gradient	$kg\,m^{-4}$
$\mathbf{grad}\,pH, \nabla pH$	pH gradient	m^{-1}
$\mathbf{grad}\,T, \nabla T$	temperature gradient	$K\,m^{-1}$
$\mathbf{grad}\,\psi, \nabla \psi_m$	massic energy gradient	$m\,s^{-2}$
G	Gibbs energy, Gibbs free energy, Gibbs function	$J = kg\,m^2\,s^{-2}$
G	gravitational constant	$N\,m^2\,kg^{-2} = m^3\,kg^{-1}\,s^{-2}$
G	electrical conductance, electric conductance	$S = kg^{-1}\,m^{-2}\,s^3\,A^2$
G	kelvic heat rate, thermal conductance, thermal conductivity, thermal conduction coefficient	$W\,K^{-1} = kg\,m^2\,s^{-3}\,K^{-1}$
G	Gibbs energy of activation	$J\,mol^{-1} = kg\,m^2\,mol^{-1}\,s^{-2}$
h, z	height	m
hv	entitic energy of photons	$J = kg\,m^2\,s^{-2}$
h	coefficient of heat transfer	$W\,m^{-2}\,K^{-1} = kg\,s^{-3}\,K^{-1}$
h	massic enthalpy, specific enthalpy	$J\,kg^{-1} = m^2\,s^{-2}$
$h, \hbar$	Planck constant	$J\,s = kg\,m^2\,s^{-1}$
H	effective massic energy of ionizing radiation	$m^2\,s^{-2}$
H	enthalpy	$J = kg\,m^2\,s^{-2}$

Symbol(s) of kind-of-quantity	Term(s) for kind(s)-of-quantity	Symbol(s) for SI base unit or coherent derived SI unit
H, $\boldsymbol{H}$, $\vec{H}$	magnetic field strength	$\mathrm{A\ m^{-1}}$
H_v, H	areic light, luminous exposure, light exposure	$\mathrm{lx\ s = lm\ m^{-2}\ s}$
I	mass of food or nutrient ingested, intake of food or nutrient	kg
I, $\boldsymbol{I}$	impulse	$\mathrm{N\ s = kg\ m\ s^{-1}}$
I, i	electrical current, electricity rate, electric current	A
I, J	moment of inertia, dynamic moment of inertia	$\mathrm{kg\ m^2}$
I, J	areic energy rate of a unidirectional sound wave, sound intensity	$\mathrm{kg\ s^{-3}}$
I, M	insulation coefficient, insulance, coefficient of thermal insulation	$\mathrm{K\ m^2\ W^{-1} = K\ s^3\ kg^{-1}}$
I_b, I_m	ionic strength (molality basis), molal ionic strength, mean ionic molality	$\mathrm{mol\ kg^{-1}}$
I_c	ionic strength (substance concentration basis), concentrational ionic strength	$\mathrm{mol\ m^{-3}}$
I_e, I	steradic energy rate of radiation, radiant intensity	$\mathrm{W = kg\ m^2\ s^{-3}}$
I_v, I	luminous intensity, steradic light rate	$\mathrm{cd\ m^{-2}}$
j, $\boldsymbol{J}$	areic electrical current, areic electricity rate, electric current density	$\mathrm{A\ m^{-2}}$
J, I	moment of inertia, dynamic moment of inertia	$\mathrm{kg\ m^2}$
J, I	areic energy rate of a unidirectional sound wave, sound intensity	$\mathrm{kg\ s^{-3}}$
$\boldsymbol{J}$, $\boldsymbol{L}$, $\vec{L}$	angular momentum, momentum of momentum	$\mathrm{kg\ m^2\ s^{-1}}$
J_m, F_m, $\dot{m}$, Φ_m	mass rate, mean mass rate, mass flow rate, mass rate of flow, mass transfer rate, mass velocity	$\mathrm{kg\ s^{-1}}$
J_n, φ_n	areic substance rate, substance flux density	$\mathrm{mol\ m^{-2}\ s^{-1}}$
J_x, J	flux of a quantity X	(varies)
J_v, φ_v	areic volume rate	$\mathrm{m\ s^{-1}}$
k	coverage factor	1
k, τ	time coefficient	s
k	rate coefficient, rate coefficient for chemical reaction	$\mathrm{s^{-1}}$
k	lineic electrical conductance, electrical conductivity	$\mathrm{S\ m^{-1} = s^3\ A^2\ kg^{-1}\ m^{-3}}$
k, l	thermal conductivity	$\mathrm{W\ m^{-1}\ K^{-1} = kg\ m\ s^{-3}\ K^{-1}}$
k_B, k	entitic kelvic energy constant, Boltzmann constant, molecular gas constant, entitic gas constant	$\mathrm{J\ K^{-1} = kg\ m^2\ s^{-2}\ K^{-1}}$
k_{elim}	rate coefficient for elimination, elimination rate coefficient	$\mathrm{s^{-1}}$
K, a_l, a	lineic decadic absorbance, linear decadic absorption coefficient	$\mathrm{m^{-1}}$
K	cubic pressure coefficient	$\mathrm{Pa = kg\ m^{-1}\ s^{-2}}$

Symbol(s) of kind-of-quantity	Term(s) for kind(s)-of-quantity	Symbol(s) for SI base unit or coherent derived SI unit
K_p	equilibrium coefficient (based on partial pressure) of chemical reaction, baric equilibrium product	$Pa = kg\ m^{-1}\ s^{-2}$
K, α	kelvic areic heat rate, heat transfer coefficient, coefficient of heat transfer	$W\ m^{-2}\ K^{-1} = kg\ s^{-3}\ K^{-1}$
K	luminous efficacy	$cd\ m^{-2}\ kg^{-1}\ s^3$
K_b, K_m	molal equilibrium coefficient, equilibrium constant (molality basis), molal equilibrium product	$mol\ kg^{-1}$
K_M	Michaelis–Menten coefficient, Michaelis–Menten constant, Michaelis constant	$mol\ m^{-3}$
K_c	substance-concentrational equilibrium coefficient of reaction, concentrational equilibrium product	$mol\ m^{-3}$
$K_{fus,c}$	substance-concentrational freezing-point coefficient, concentrational freezing point depression constant	$m^3\ K\ mol^{-1}$
$K_{fus,b}$	molal freezing-point coefficient, molal freezing point depression constant	$kg\ K\ mol^{-1}$
l, L	length, distance, length traversed	m
l, k	thermal conductivity	$W\ m^{-1}\ K^{-1} = kg\ m\ s^{-3}\ K^{-1}$
L	self-inductance	$H = kg\ m^2\ s^{-2}\ A^{-2}$
L, N_A	Avogadro constant	mol^{-1}
$L_{e,\lambda}, L_\lambda$	differential quotient of steradic areic energy rate to wavelength	$kg\ m^{-1}\ s^{-3}$
L_p	hydraulic permeability	$m\ Pa^{-1}\ s^{-1} = s\ m^2\ kg^{-1}$
$\boldsymbol{L}, \vec{L}, \boldsymbol{J}$	angular momentum, momentum of momentum	$kg\ m^2\ s^{-1}$
m	mass, atomic mass, molecular mass, entitic mass of entities, entitic mass of atoms, atomic mass	kg
m, b	molality	$mol\ kg^{-1}$
m_A, ρ_A	areic mass	$kg\ m^{-2}$
m_l, ρ_l	lineic mass	$kg\ m^{-1}$
m_r	mass ratio	1
m_s	standard mass	kg
$m(\lambda)$	lineic attenuance to radiation	m^{-1}
$\dot{m}, \Phi_m, F_m, J_m$	mass rate, mean mass rate, mass flow rate, mass rate of flow, mass transfer rate, mass velocity	$kg\ s^{-1}$
M	mass, mass of individual, mass of particle, rest mass	kg
M_r	relative molar mass	1
M, M_e	areic energy rate of radiation emitted, radiant exitance, brightness, irradiance, radiant emittance	$kg\ s^{-3}$
M	molar mass	$kg\ mol^{-1}$
M, I	insulation coefficient, insulance, thermal insulance, coefficient of thermal insulation	$K\ m^2\ W^{-1} = K\ s^3\ kg^{-1}$
M_v	illuminance, areic light rate	$cd\ sr\ m^{-2}$
$M, \boldsymbol{M}, \vec{M}$	moment of force	$J = kg\ m^2\ s^{-2}$

Symbol(s) of kind-of-quantity	Term(s) for kind(s)-of-quantity	Symbol(s) for SI base unit or coherent derived SI unit
n	number of entities, absolute frequency	1
n, z	charge number of a cell reaction, partial order of reaction	1
n	relative lineic time for radiation, refractive index	1
n	relative amount-of-substance	1
n	partial order of reaction	1
n	volumic number	m^{-3}
n	dynamic viscosity, coefficient of internal friction	$Pa\ s = kg\ m^{-1}\ s^{-1}$
n	amount-of-substance, molar amount	mol
n_A	areic substance	$mol\ m^{-2}$
$\dot{n}, F_n, \Phi_n$	substance rate, substance flow rate	$mol\ s^{-1}$
N	number of entities, absolute frequency	1
N	number content	kg^{-1}
N_A, L	Avogadro constant	mol^{-1}
N_A	areic number	m^{-2}
N_r	number ratio	1
$\dot{N}, \nu, \Phi_N, f$	number rate	s^{-1}
p, P	number fraction expected, probability	1
$p, \boldsymbol{p}$	momentum	$N\ s = kg\ m\ s^{-1}$
p	pressure, absolute pressure, partial pressure, tension of gas, gas tension	$Pa = kg\ m^{-1}\ s^{-2}$
$\tilde{p}, f$	active pressure, fugacity	$Pa = kg\ m^{-1}\ s^{-2}$
p	electric resistivity	$\omega\ m = kg\ m^3\ a^{-2}\ s^{-3}$
p_s	static pressure	$Pa = kg\ m^{-1}\ s^{-2}$
ρ, γ	mass concentration	$kg\ m^{-3}$
pH	pH, hydrogen ion exponent	1
pH(I)	isoelectric point	1
P, p	number fraction expected, probability	1
P, Φ_e	energy rate, power	$W = kg\ m^2\ s^{-3}$
$P(\Delta\lambda), \Phi_e(\Delta\lambda)$	energy rate of radiation, radiant flux, radiant power, radiant energy flux	$W = kg\ m^2\ s^{-3}$
$P_\nu, \Phi_{e,\nu}, \Phi_\nu$	differential quotient of energy rate to frequency	$W\ Hz^{-1} = kg\ m^3\ s^{-3}$
q	massic electrical charge	$C\ kg^{-1} = A\ s\ kg^{-1}$
q, Q	amount-of-heat, quantity of heat	$J = kg\ m^2\ s^{-2}$
Q, E	energy, amount-of-energy	$J = kg\ m^2\ s^{-2}$
Q	electrical charge, electric charge, quantity of electricity, electrical amount, amount-of-electricity	$C = A\ s$
Q, q	amount-of-heat, quantity of heat	$J = kg\ m^2\ s^{-2}$
Q, W	radiant energy	$J = kg\ m^2\ s^{-2}$
Q, Q_v	amount-of-light, luminous amount, quantity of light	$lm\ s$
Q_e	energy of radiation, radiant energy	$J = kg\ m^2\ s^{-2}$
$Q_{e,m}$	molar energy of photons	$J\ mol^{-1} = kg\ m^2\ mol^{-1}\ s^{-2}$
r, d	distance travelled or migrated	m
r, R	radius of circle, centrifugal radius of circle	m

Symbol(s) of kind-of-quantity	Term(s) for kind(s)-of-quantity	Symbol(s) for SI base unit or coherent derived SI unit
r	rate of concentration change	$\mathrm{mol\ m^{-3}\ s^{-1}}$
$r_{B/C}$	substance ratio of component B to component C, mole ratio, amount-of-substance ratio	1
R	electric resistance	$\Omega = \mathrm{kg\ m^2\ s^{-3}\ A^{-2}}$
R, ρ	reflectance	1
R	thermal resistance	$\mathrm{K\ W^{-1}} = \mathrm{K\ s^3\ kg^{-1}\ m^{-2}}$
R	molar kelvic energy constant, molar gas constant	$\mathrm{J\ K^{-1}\ mol^{-1}}$ $= \mathrm{kg\ m^2\ s^{-2}\ K^{-1}\ mol^{-1}}$
s, S	relative amount-of-substance	1
s	length of path travelled	m
s	sedimentation coefficient	s
s	massic entropy	$\mathrm{J\ kg^{-1}\ K^{-1}} = \mathrm{m^2\ s^{-2}\ K^{-1}}$
S	area	$\mathrm{m^2}$
S	entropy	$\mathrm{J\ K^{-1}} = \mathrm{kg\ m^2\ s^{-2}\ K^{-1}}$
t	internal transmittance	1
t, T	transmittance	1
t	time, time duration, time elapsed, time of centrifugation	s
t, θ	Celsius temperature	$\mathrm{^\circ C = K}$
$t_{1/2}, T_{1/2}, T_{0.5}$	half-life, half-value time, half-time, half-disappearance time	s
t_r	relative time	1
T	entitic time	s
T, Φ	thermodynamic temperature, Kelvin temperature, absolute temperature	K
T, E_k, E_{kin}	kinetic energy	$\mathrm{J} = \mathrm{kg\ m^2\ s^{-2}}$
$T_{1/2}, t_{1/2}, T_{0.5}$	half-life, half-value time, half-time	s
$\dot{T}$	temperature rate	$\mathrm{K\ s^{-1}}$
$\boldsymbol{u, v, w, c}$	length rate vector, velocity	$\mathrm{m\ s^{-1}}$
u, v	speed	$\mathrm{m\ s^{-1}}$
u	massic thermodynamic energy, massic internal energy	$\mathrm{J\ kg^{-1}} = \mathrm{m^2\ s^{-2}}$
u_c	combined standard measurement uncertainty	varies
U	expanded measurement uncertainty	varies
U	thermodynamic energy, internal energy	$\mathrm{J} = \mathrm{kg\ m^2\ s^{-2}}$
U_m	molar thermodynamic energy, molar internal energy	$\mathrm{J\ mol^{-1}} = \mathrm{kg\ m^2\ mol^{-1}\ s^{-2}}$
$U, \Delta V$	electric potential difference	$\mathrm{V} = \mathrm{kg\ m^2\ s^{-3}\ A^{-1}}$
ν	degree of freedom	1
ν	length rate, speed, rate of migration, rate of travel, velocity of migration, sedimentation velocity	$\mathrm{m\ s^{-1}}$
$\boldsymbol{v, u, w, c}$	length rate vector, rate of travel, velocity	$\mathrm{m\ s^{-1}}$
v	massic volume, partial massic volume	$\mathrm{m^3\ kg^{-1}}$
v	volume content	$\mathrm{m^3\ kg^{-1}}$
$\nu, \Phi_N, f, \dot{N}$	number rate	$\mathrm{s^{-1}}$
ν, f	number rate of regular event, frequency	$\mathrm{Hz} = \mathrm{s^{-1}}$

Symbol(s) of kind-of-quantity	Term(s) for kind(s)-of-quantity	Symbol(s) for SI base unit or coherent derived SI unit
v	substance concentration rate of reaction (based on amount concentration), rate of reaction	$mol\ m^{-3}\ s^{-1}$
v	substance content rate	$mol\ kg^{-1}\ s^{-1}$
V	volume	m^3
V, φ	electric potential	$V = kg\ m^2\ s^{-3}\ A^{-1}$
$\dot{V}, \Phi_v, F_v$	volume rate, volume flow rate, mean volume rate, volume rate of flow	$m^3\ s^{-1}$
$\dot{V}, \varphi_v$	areic volume rate	$m\ s^{-1}$
V_m	molar volume, partial molar volume	$m^3\ mol^{-1}$
V_r	volume ratio	1
w	mass fraction	1
$\boldsymbol{w, u, v, c}$	length rate vector, velocity	$m\ s^{-1}$
w	volumic energy	$J\ m^{-3} = kg\ m^{-1}\ s^{-2}$
w_e	volumic energy rate	$kg\ m^{-1}\ s^{-3}$
w, W	work	$J = kg\ m^2\ s^{-2}$
x, y	amount-of-substance fraction, substance fraction, mole fraction	1
x	rate of conversion	$mol\ s^{-1}$
X	massic electrical charge induced by ionizing radiation, exposure	$C\ kg^{-1} = A\ s\ kg^{-1}$
y, b	breadth	m
y, x	amount-of-substance fraction, substance fraction, mole fraction	1
z	charge number	1
z, h	height	m
z	electrokinetic potential	$V = kg\ m^2\ s^{-3}\ A^{-1}$
$-z, d, \delta$	depth	m
z_f	catalytic activity fraction	1
z_E	catalytic activity, enzyme activity, enzym(at)ic activity, catalytic amount	$mol\ s^{-1}$
$\dot{z}_E$	catalytic-activity rate	$kat\ s^{-1}$
Z	impedance, complex impedance	$\Omega = kg\ m^2\ s^{-3}\ A^{-2}$
Z_m	mechanical impedance	$N\ s\ m^{-1} = kg\ s^{-1}$
$Z_{m,E}$	molar catalytic activity of enzyme	$kat\ mol^{-1}$
α	absorptance, internal absorptance	1
α	expansion coefficient	1
α	number fraction of molecules dissociated, degree of dissociation, dissociation fraction	1
α	rate coefficient for elimination, elimination rate constant	s^{-1}
α	angular acceleration	s^{-2}
$\boldsymbol{\alpha}, \gamma$	electrical polarizability	$C\ m^2\ V^{-1} = A^2\ s^4\ kg^{-1}$
α, K	kelvic areic heat rate, heat transfer coefficient, coefficient of heat transfer	$W\ m^{-2}\ K^{-1} = kg\ s^{-3}\ K^{-1}$
α_b	molal solubility coefficient	$mol\ kg^{-1}\ Pa^{-1}$ $= mol\ m\ s^2\ kg^{-2}$
α_c	substance-concentrational solubility coefficient	$mol\ s^2\ kg^{-1}\ m^{-2}$

Symbol(s) of kind-of-quantity	Term(s) for kind(s)-of-quantity	Symbol(s) for SI base unit or coherent derived SI unit
α_l	lineic napierian absorbance, linear napierian absorption coefficient, linear napierian absorptivity	m^{-1}
α_l, α, K	lineic decadic absorbance, linear decadic absorption coefficient	m^{-1}
α_l	linear expansion coefficient	$m\ K^{-1}$
α_p	relative pressure coefficient, pressure coefficient	$K^{-1} = C^{-1}$
α_p	massic area rotance, massic optical rotatory power	$m^2\ kg^{-1}$
α_s	specific napierian attenuation coefficient, specific napierian extinction coefficient	1
α_V	cubic expansion coefficient	$m^3\ K^{-1}$
α_x	substance-fractional solubility coefficient, rational solubility coefficient	$Pa^{-1} = m\ s^2\ kg^{-1}$
α_m	molar area	$m^2\ mol^{-1}$
β	kelvic pressure	$Pa\ K^{-1} = kg\ m^{-1}\ s^{-2}\ K^{-1}$
β	differential quotient of stoichiometric substance concentration of hydrogen ion to pH, volumic buffer capacity	$mol\ m^{-3}$
γ	activity factor, activity coefficient, substance-concentrational activity factor substance-concentrational activity coefficient	1
γ, α	electrical polarizability	$C\ m^2\ V^{-1} = A^2\ s^4\ kg^{-1}$
γ, ρ	volumic mass, partial volumic mass	$kg\ m^{-3}$
γ, σ	areic energy of interface, surface tension	$J\ m^{-2} = kg\ s^{-2}$
γ, σ, κ	lineic electrical conductance, electrical conductivity, electrolytic conductivity	$S\ m^{-1} = A^2\ s^4\ kg^{-1}\ m^{-1}$
γ_b, γ_m	molal activity factor, molal activity coefficient	1
γ_x, f_x	substance-fractional activity factor, rational activity coefficient	1
δ	number fraction, particle fraction, entity fraction	1
δ, d, $-z$	depth	m
δ, λ	rate coefficient	s^{-1}
δ_{dis}	number fraction of people, prevalence	1
δ_{inf}	number fraction of people infected, prevalence of a given infection, proportion of people infected	1
ε	molar area decadic absorbance, molar linear decadic absorption coefficient, molar decadic absorption coefficient, molar decadic absorptivity	$m^2\ mol^{-1}$
ζ	electrokinetic potential, zeta potential	$V = kg\ m^2\ s^{-3}\ A^{-1}$
η	dynamic viscosity, coefficient of internal friction	$Pa\ s = kg\ m^{-1}\ s^{-1}$
θ, t	Celsius temperature	$°C = K$

Symbol(s) of kind-of-quantity	Term(s) for kind(s)-of-quantity	Symbol(s) for SI base unit or coherent derived SI unit
κ	attenuation coefficient	1
κ	molar area naperian absorbance, molar linear naperian coefficient, molar naperian absorption coefficient, molar naperian absorptivity	$m^2\ mol^{-1}$
κ	conductivity	$S\ m^{-1} = A^2\ s^4\ kg^{-1}\ m^{-1}$
κ, λ	lineic kelvic heat rate, thermal conductivity, thermal conduction coefficient	$W\ m^{-1}\ K^{-1}$ $= kg\ m\ s^{-3}\ K^{-1}$
κ, γ, σ	lineic electrical conductance, electrical conductivity, electrolytic conductivity	$S\ m^{-1} = A^2\ s^4\ kg^{-1}\ m^{-1}$
κ_E, b_E, e_E	catalytic-activity concentration, catalytic concentration	$kat\ m^{-3} = mol\ m^{-3}\ s^{-1}$
λ	chemical activity, absolute chemical activity	1
λ	wavelength	m
λ	decay coefficient for radioactivity, disintegration coefficient	s^{-1}
λ, Λ_m	molar area electrical conductance, molar electrical conductivity, molar conductivity, ionic conductivity	$S\ m^2\ mol^{-1}$
λ, σ, κ	lineic electrical conductance, electrical conductivity, electrolytic conductivity	$S\ m^{-1} = A^2\ s^4\ kg^{-1}\ m^{-1}$
λ, κ	lineic kelvic heat rate, thermal conductivity, thermal conduction coefficient	$W\ m^{-1}\ K^{-1} = kg\ s^{-3}\ K^{-1}$
μ, μ_l	linear napieran attenuation coefficient	m^{-1}
μ	electrophoretic mobility, electrical mobility, electrolytic mobility	$A\ s^2\ kg^{-1} = m^2\ s^{-1}\ V^{-1}$
μ	chemical potential, absolute chemical potential, electrochemical potential, absolute electrochemical potential	$J\ mol^{-1} = kg\ m^2\ mol^{-1}\ s^{-2}$
ν	stoichiometric number	1
ν	kinematic viscosity	$m^2\ s^{-1}$
ν	volume content, massic volume, partial massic volume	$m^3\ kg^{-1}$
ν, f	number rate of regular event, frequency	$Hz = s^{-1}$
ν	substance content	$mol\ kg^{-1}$
ξ	linear displacement	m
Π, ψ	osmotic pressure, osmotic volumic energy	$Pa = m\ s^2\ kg^{-1}$
ρ, R	reflectance	1
ρ	radius of curvature of arc	m
ρ, γ	mass concentration, volumic mass, partial volumic mass	$kg\ m^{-3}$
ρ	mass concentration rate	$kg\ m^{-3}\ s^{-1}$
ρ_r	relative mass concentration	1
σ	lineic number, repetency, lineic number of waves, wavenumber	m^{-1}
σ	normal stress	$Pa = kg\ m^{-1}\ s^{-2}$

Symbol(s) of kind-of-quantity	Term(s) for kind(s)-of-quantity	Symbol(s) for SI base unit or coherent derived SI unit
σ, γ, κ	lineic electrical conductance, electrical conductivity, electrolytic conductivity	$S\ m^{-1} = A^2\ s^4\ kg^{-1}\ m^{-1}$
σ, γ	areic energy of interface, surface tension	$N\ m^{-1} = kg\ s^{-2}$
τ	shear stress	$Pa = kg\ m^{-1}\ s^{-2}$
τ, k	time coefficient	s
φ, ϑ	volume fraction	1
φ, Φ	quantum yield, photochemical yield	1
Φ	osmotic coefficient, osmotic factor	1
Φ	areic heat rate, areic heat flow rate	$J\ m^{-2}\ s^{-1}$
φ, V	electric potential	$V = kg\ m^2\ s^{-3}\ A^{-1}$
φ_b, φ_m	molal osmotic factor, molal osmotic coefficient	1
φ_c	substance-concentrational osmotic factor, concentrational osmotic coefficient	1
φ_m	areic mass rate	$kg\ m^{-2}\ s^{-1}$
φ_n, J_n	areic substance rate, substance flux density	$mol\ m^{-2}\ s^{-1}$
$\Phi_N, f, \nu, \dot{N}$	number rate	s^{-1}
$\varphi_V, J_V, \dot{V}$	areic volume rate	$m\ s^{-1}$
φ_x	substance-fractional osmotic coefficient, substance-fractional osmotic factor	1
Φ	heat rate, heat flow rate	$W = kg\ m^2\ s^{-3}$
Φ, T	thermodynamic temperature, Kelvin temperature, absolute temperature	K
Φ_e, P	energy rate, power	$W = kg\ m^2\ s^{-3}$
$\Phi_e(\Delta\lambda), P(\Delta\lambda)$	energy rate of radiation, radiant flux, radiant power, radiant energy flux	$W = kg\ m^2\ s^{-3}$
$\Phi_{e,\lambda}, P_\lambda$	differential quotient of energy rate to wavelength	$W\ m^{-1}$
$\Phi_{e,\tilde{\nu}}, P_{\tilde{\nu}}$	differential quotient of energy rate to wavenumber, spectral concentration of radiant power in terms of wavenumber, spectral radiant power in terms of wavenumber	$W\ m^{-1}$
Φ_m	magnetic flux	$Wb = kg\ m^2\ s^{-2}\ A^{-1}$
$\Phi_m, \dot{m}, F_m, J_m$	mass rate, mean mass rate, mass flow rate, mass rate of flow, mass transfer rate, mass velocity	$kg\ s^{-1}$
$\Phi_n, F_n, \dot{n}$	substance rate, substance rate of reaction, substance flow rate	$mol\ s^{-1}$
$\Phi_v, F_v, \dot{V}$	volume rate, volume flow rate, volume rate of flow, mean volume rate	$m^3\ s^{-1}$
Φ_v, Φ	light rate, luminous flux	lx
Φ_V, J_V	areic volume rate	$m\ s^{-1}$
$\boldsymbol{\Phi_{N,\Phi}}$	number rate of photons	s^{-1}
ψ	areic energy, energy fluence	$J\ m^{-2} = kg\ s^{-2}$
Ψ, Π	osmotic pressure, osmotic volumic energy	$Pa = kg\ m^{-1}\ s^{-2}$
ω	angle rate, angular velocity, circular frequency, pulsatance, angular frequency	s^{-1}, $rad\ s^{-1}$

Structure of Numbered Entries in Sections 8 and 9

For kinds-of-quantity

⟨2⟩ systematic term(s)
⟨3⟩ synonym(s)
⟨4⟩ abbreviation
⟨5⟩ preferred symbol(s)
⟨6⟩ definition in words
⟨7⟩ definition by equation
⟨8⟩ deprecated term(s) or symbol(s)
⟨9⟩ note(s)
⟨10⟩ example(s) of dedicated kinds-of-quantity

bold characters: recommended terms

For Units

⟨12⟩ term(s) for coherent SI unit(s)
⟨16⟩ definition(s) of base SI unit(s)
⟨19⟩ note(s) on use and usage of unit(s)

Synonymy

Bold type for preferred term(s)
  and acceptable synonym(s)
  italic type for other term(s)

Recommendations

Shall is used for official recommendations
  (IUPAC, IFCC, ISO, CGPM)
Should is used for preferences
  † for discouraged symbols or abbreviations

Kinds-of-quantity are listed in sections 8 and 9
Kinds-of-property without dimensions of the ISQ
  are listed in section 10

Compendium of Terminology and Nomenclature of Properties in Clinical Laboratory Sciences:
Recommendations 2016
Edited by Georges Férard, René Dybkaer and Xavier Fuentes-Arderiu
© International Union of Pure and Applied Chemistry 2017
Published by the Royal Society of Chemistry, www.rsc.org

Contents

Compendium of Terminology and Nomenclature of Properties in Clinical Laboratory Sciences:
Recommendations 2016
Edited by Georges Férard, René Dybkaer and Xavier Fuentes-Arderiu
© International Union of Pure and Applied Chemistry 2017
Published by the Royal Society of Chemistry, www.rsc.org

SECTION 1

History of Recommendations on Properties in Clinical Laboratory Sciences

The beginnings of the work on quantities and units in clinical laboratory sciences illustrate the catalytic interplay of international organizations, national societies and critical individuals with novel ideas.

In 1954, the 10th General Conference on Weights and Measures (CGPM) met and resolved that: "In accordance with the wish expressed by the 9th CGPM (of 1948) in its Resolution 6 concerning the establishment of a practical system of units of measure for international use, the 10th CGPM decides to adopt as base units ... metre, kilogram, second, ampere, degree Kelvin [renamed kelvin in 1962], candela." In 1960, that system became known as the International System of Units, SI. In 1971, the mole was added to the system on the advice of IUPAC, IUPAP, and ISO.

In the late 1950s, probably most chemists and certainly most scientists in biologically oriented disciplines dismissed the recommendations on weights and measures as irrelevant, of concern only to physicists and engineers. As parts of ISO Recommendation 31 (the forerunner of ISO 31:1992[1]) on quantities and units were being published up to 1960 on 'periodic and related phenomena', 'mechanics' and 'heat', the opinion of biologists about its irrelevance would probably have been confirmed. The titles suggested that such recommendations were not interdisciplinary.

In 1952, a IUPAC Commission on Clinical Chemistry had been formed to coordinate national groups. A professor from Geneva, Manuel C. Sanz, predecessor at the University of Geneva of another active worker for IFCC, Marc Roth, encountered a group of young medical graduates during a visit to Copenhagen. The group was pursuing further studies in biochemistry, including analytical, inorganic, organic and physical chemistry. Sanz became aware of the local ideas on working towards a unified "presentation of results" in clinical chemistry and took immediate steps to bring their ideas to the attention of the international scientific community.

In 1957, René Dybkær had been invited by the Danish Society for Clinical Chemistry and Clinical Physiology to give a paper entitled "Standardisering af enheds-betegnelser i klinisk-kemisk laboratoriearbejde" (Standardizing unit symbols in work of the clinico-chemical laboratory). The paper was subsequently published in the journal *Nordisk Medicin*.[2] It was a plea for implementation in clinical chemistry of the recommendations of the IUPAC Commission on Symbols and Physicochemical Terminology and of ISO Technical Committee 12 on quantities and units.

Compendium of Terminology and Nomenclature of Properties in Clinical Laboratory Sciences: Recommendations 2016
Edited by Georges Férard, René Dybkaer and Xavier Fuentes-Arderiu
© International Union of Pure and Applied Chemistry 2017
Published by the Royal Society of Chemistry, www.rsc.org

The strength of the beginnings in Denmark was the realization that the recommendations of BIPM and ISO, despite their orientation to physical sciences, had an interdisciplinary element that needed to be sought out and applied. The Danish society appointed a committee (R. Dybkær, N. S. C. Heilskov, J. Jørgensen, K. Jørgensen, and E. Praetorius), which presented a proposal in 1960 to the other Scandinavian societies of clinical chemistry and also translated it as a proposal to the Fourth International Congress on Clinical Chemistry in Edinburgh in 1960. The IUPAC Commission on Clinical Chemistry referred the proposal to the Fifth International Congress on Clinical Chemistry in Detroit in August 1963. Meanwhile, ideas were developing further in Scandinavia. The relevance of molecular quantities (§7.1.2), although not then part of the International System of Units, was stressed.

The Committee on Standards and Control of the American Association of Clinical Chemistry discussed the proposal in 1963 and its comments formed part of the subsequent recommendation. The situation as seen from the United States is described by N. Radin:[3] "A system of units for clinical chemists was proposed by the Danish Society for Clinical Chemistry and Clinical Physiology. This system was transmitted to the IUPAC Commission on Clinical Chemistry in June 1963. A Subcommittee on Nomenclature and Usage of the Committee on Standards and Controls of the American Association of Clinical Chemistry was assigned the task of studying the problem with a view to agreement upon a set of rational and internally consistent objectives for nomenclature reform. The recommendations of the subcommittee were in close agreement with those of the original Danish proposals as being desirable objectives."

The proposal was accepted as a basis for a recommendation at the Fifth International Congress (1963), where a symposium was held on 'Nomenclature and standard units'. The draft recommendation, prepared by Dybkær and Jørgensen, was presented to the IUPAC Commission on Clinical Chemistry in 1964.

At the IUPAC General Assembly in Paris in 1965, the Commission resolved to recommend the cubic metre as the preferred denominator unit in expression of concentrations (§5.10.2) and appointed a Subcommission for Standards and Units in Clinical Chemistry consisting of Jean-Emile Courtois (Paris), René Dybkær, Per Lous (Copenhagen), Noel F. Maclagan (London), Martin Rubin (Washington, DC) and Manuel C. Sanz. At the first meeting of the Subcommission in Copenhagen in November 1965, Poul Astrup and Kjeld Jørgensen attended as 'experts'. The draft, based on 'A primer of quantities and units in clinical chemistry', 1966, was finalized and approved by the IUPAC Commission on Clinical Chemistry and by the Council of IFCC during the Sixth International Congress of Clinical Chemistry in Munich in 1966. It was included as a Recommendation 1966 in a monograph by Dybkær and Jørgensen, and published in 1967 under the title "Quantities and units in clinical chemistry".[4]

The change in title and in emphasis is remarkable. Superficially the problem seemed to be units and their representation. Evolution from 1957 to 1967 perhaps reflects the realization that a unified way of naming and specifying the quantities was more fundamental; standardization of measurement units was a secondary issue.

The principle of preference for amount-of-substance rather than mass in describing chemical amounts was approved in 1966 by the clinical chemists and ratified in 1972 by the International Committee (now Council) for Standardization in Haematology (ICSH), IFCC and the World Association of Societies of Anatomic and Clinical Pathology (WASP). The recommendation was published in 1973,[5] together with one on preference for the litre in expression of concentrations, under the title "Use of SI in clinical laboratory measurements". Concurrently, the mole had been accepted by CGPM in 1971 as an SI base unit for amount of substance.

Two proposals in the 1966 primer failed to reach approval: 'Whole number measurement of quantities' and 'Qualitative measurement of quantities'. These issues have continued to plague the reporting of analytical data and have surfaced again in discussions in recent years (§4.1).

In 1971, the IUPAC Commission and IFCC Expert Panel (which became a Committee in the 1980s) published two of the most useful guides on the reporting of biological and chemical data, the IUPAC 'Yellow Appendices' Nos 20 and 21 of 1972,[6,7] which were provisional abridged revisions of the 1967 Recommendations. Revision of them started during the IUPAC General Assembly at Munich in 1973.

The group met at Munich again in 1974: René Dybkær, Bernard H. Armbrecht from the US Food & Drug Administration in Washington, DC, Kjeld Jørgensen, Pierre Métais from the University of Strasbourg, Roland Herrmann from the University of Giessen and J. Christopher Rigg from Wageningen. Besides the revision of the 1967 publication, of which the Yellow Appendices Nos. 20 and 21 were the first stage, drafts were discussed on fluid mechanics (little of which has seen the light of day), on spectroscopy (initiated by Métais and Herrmann) and on activities and pH. The discussions were extremely thorough.

In contrast to the rapid progress towards agreement among clinical laboratory scientists on ways of expressing concentrations, the problems of expressing enzymological data (§7.6) were just being realized for the first time. Dybkær had a seemingly endless itinerary in the early 1970s discussing enzyme activity and catalytic amount and their units, the enzyme unit and the katal, with ISO Technical Committee 12, IUPAC, the International Union of Biochemistry (IUB) Joint Committee on Biochemical Nomenclature, and the authorities of IUPAC and IFCC. The problems were resolved by the IUPAC Commission on Quantities and Units in Clinical Chemistry (CQUCC), the IFCC Expert Panel on Quantities and Units (EPQU) and the IFCC Expert Panel on Enzymes at a joint meeting at Strasbourg in 1975 chaired by Donald Moss. The results of the negotiations with the IUPAC–IUB Joint Committee on Biochemical Nomenclature were published in 1978. A revised version of the recommendation of the commission and expert panel was at last published in 1979. The IFCC, through IUPAC, had asked the Consultative Committee for Units (CCU) of the International Committee for Weights and Measures (CIPM) to support a petition to the CGPM for a special term "katal" for the SI coherent unit 'mole per second' in expressing 'catalytic activity'. The CCU was not convinced of the need, although in the same period radiologists managed to win approval of special terms for the derived units becquerel, gray and sievert (§5.8.2, Table 5.2) with similar arguments about human safety.

At the IUPAC General Assembly at Madrid in September 1975, there were joint meetings with the two spectroscopic commissions of IUPAC, who (quite rightly) tore to pieces an early draft on quantities and units in spectroscopy: current nomenclature was a mess, but logical and user-friendly answers were still needed. The discussion with the physical chemists yielded permission to introduce the term **substance concentration** instead of the long-winded **amount-of-substance concentration**. However, discussion on whether the unit of catalytic amount, termed the katal, was a base unit or a derived unit was extremely irate, with Max L. McGlashan, the then chairman of the IUPAC Interdivisional Committee on Nomenclature and Symbols (IDCNS) addressing the members of the CQUCC in hortatory fashion as "naughty" for trying to introduce new nomenclature without proper discussion. The members of the Commission were sent away with instructions to introduce the new nomenclature properly (§5.4). At times, that project was extremely frustrating, despite the encouragement of a subsequent chairman of IDCNS, Norman Jones, who prophesied that it would take at least another 10 years.

It is gratifying to note that when IFCC approached the CIPM in 1998 with the proposal for the special term "katal", it was this time supported by CCU and the Consultative Committee for Amount of Substance (CCQM). This led to acceptance by Resolution 12 of the 21st CGPM in 1999, meaning that katal is a *bona fide* SI unit term in the relevant circumstances.[8]

In the 1970s, metrologists and the physical chemists at Madrid looked askance at the distinction Dybkaer and Jørgensen had made in 1967 between quantity and kind-of-quantity. By contrast, three past members of IUPAC–CQUCC and IFCC–EPQU participated in the committee of an ISO task group responsible for producing the second edition of the 'International Vocabulary of Metrology (VIM)'[9] and clinical chemists also sat in the group that made significant changes to the third edition,[10] including adoption of the concept 'kind of quantity'. Clinical laboratory sciences are now being taken seriously as a branch of metrology.

The need for clarity of intergovernmental policy on SI in the medical professions led to the preparation by WHO of 'A guide to international recommendations on names and symbols for quantities and on units of measurement',[11] in which the group provided considerable information about IUPAC and IFCC recommendations. This led on to a resolution of the 30th World Health Assembly (WHA) at Geneva in May 1977. The group, in particular its chairman Robert Zender (La Chaux-de-Fonds, Switzerland), advised on the formulation of the recommendation and on the subsequent publication by WHO[12] of a booklet on 'The SI for the health professions', as requested by the 30th WHA. There had been strong requests, notably from cardiologists, for clarification of policy on units for blood pressure and that issue was clarified for the time being in a resolution of the 34th WHA in 1981 (§9.41U, Note 2).

A continuing issue has been the metrological status of WHO International Units (IU) (§10.3) and the problems of using biological reference materials (§6.11.6) as standards in biological, immunological and chemical measurements.

Introduction of the general language of metrology into a biologically oriented discipline has been a two-way process. One example was the deficiencies found in the terms of spectrometric quantities. The resulting thorough critique seriously delayed the publication of a recommendation on absorption spectrometry, which was finalized jointly with the IUPAC Commission on Molecular Structure and Spectroscopy.[13] However, it did force the clinical chemists to initiate a separate project on nomenclature of derived quantities, which was finalized by H. Peter Lehmann (Louisiana State University), who took over as chairman in 1979.[14] That recommendation considerably influenced the nomenclature of ISO 31:1992.[1]

Some of the ideas debated in the commission and expert panel have proved too contentious or too theoretical for publication as recommendations. After Robert Zender took over as chairman of the group from Dybkær in 1975, his attempt to tighten up the definitions of basic concepts such as system and component in terms of set theory foundered because they were felt to be too theoretical. Some of his ideas have been published[15] and have been taken account of in drafting §4.3.1 and 5.1. Section 5.1 also takes account of discussions in IFCC on 'semiquantitative' and 'qualitative tests', which were published as position papers by Dybkær and Jørgensen[16] and by Dybkaer[17] who pursued the matter further in a thesis in 2004 and an updated text in 2009.[18]

Another area where the group has done pioneering work, although unpublishable as recommendations, was logarithmic quantities (§8.10), which were examined by Ben Visser (Maastricht, the Netherlands). Some of his ideas were published in papers of the IFCC Expert Panel on pH and Blood Gases.[19] His wisdom on the wider issues of the unit one[19] has not yet been fully exploited (§5.13), although it has influenced the revision of ISO 31-6 on acoustic quantities and units.[20]

Less contentious was the work on physicochemical quantities (§7.2), including chemical activity and pH, which was later organized jointly with the IFCC Committee on pH under the chairmanship of Ole Siggaard-Andersen, also from Copenhagen. Nevertheless, symbolic notation in that area raised objections on aesthetic grounds.

At the IUPAC General Assembly in Hamburg in 1991, it was suggested that the time was ripe to consolidate the recommendations from the Division of Clinical Chemistry. Several of the initial members of the group on quantities and units in clinical chemistry were still active and could still be consulted. Several of a new generation of recommendations were near to finalization. Two key members of the group from the early 1970s had died: Roland Herrmann, who initiated the work on spectrometric quantities, in 1980; Pierre Métais, who did much to standardize nomenclature and use of quantities in the French-speaking world, in November 1987.

The task was completed by the publication in 1995 of the IFCC–IUPAC Recommendations 'Compendium of Terminology and Nomenclature of Properties in Clinical Laboratory Sciences'.[21]

Developments in metrological terminology and the necessity to expand the subject field from quantity to property, thus including nominal properties, as well as comments received on the first edition, has led to this second edition, prepared as a joint project under the aegis of the IFCC Committee and IUPAC Subcommittee on Nomenclature for Properties and Units.

The work on a coding scheme for clinical laboratories was initiated in 1987 by Henrik Olesen (Copenhagen), newly elected chairman of the Commission on Quantities and Units (C-QU) of IFCC and IUPAC. With Olesen as convener of Working Group 2 on Healthcare Terminology, Semantics and Knowledge Bases of the Technical Committee 251 on Standards in Healthcare Informatics and Telemetric of European Committee for Standardisation (CEN/TC251/WG2), *inter alia*, two European Norms[22,23] were elaborated that closely followed the guidelines worked out by C-QU. These standards have been combined with rules for the use of international nomenclature, *e.g.* CAS codes and terms. This has resulted in the **NPU terminology**, so far comprising definitions of more than 16 000 dedicated kinds-of-property for use in laboratory data transmission systems.[24,25] Since 2012, IFCC has published the terminology, which is updated regularly. It has been elaborated in cooperation with relevant scientific organizations in each of the subjects: properties and units in thrombosis and haemostasis, IOC prohibited drugs, clinical microbiology, trace elements, general clinical chemistry, clinical pharmacology and toxicology, reproduction and fertility, clinical allergology, clinical molecular biology, transfusion medicine and immunohaematology, and environmental human toxicology.

A history is not complete without a word about the correspondents from many parts of the world, some irate or emotional, others coldly logical. There were hundreds of comments on the various terminological documents from individual scientists and from manufacturers as well as from the following organizations:

- other groups within IUPAC and IFCC, in particular the IUPAC Interdivisional Committee on Nomenclature and Symbols, Commission I.1 on Physicochemical Symbols, Terminology and Units and the two commissions on spectroscopy;
- national societies of clinical chemistry in particular the American Association for Clinical Chemistry;
- sister scientific unions or medical societies, in particular the International Union of Biochemistry and Molecular Biology (IUBMB, formerly termed the International Union of Biochemistry, IUB), the International Council (formerly Committee) for Standardization in Haematology (ICSH) and the World Association of Societies of Anatomic and Clinical Pathology (WASP);
- intergovernmental organizations, including the World Health Organization (WHO), the International Organization for Standardization (ISO), the International Bureau of Weights and Measures (BIPM) and the International Organization for Legal Metrology (OIML); and
- regional organizations, for instance the Community Bureau of Reference (BCR), later termed the Standards Measurements and Testing Programme of the European Union, and now the Institute of Reference Materials and Measurements (IRMM).

They have all played a role in the evolution of these recommendations.

References

1. ISO 31:1992 (14 parts). Quantities and units, International Organization for Standardization, Geneva. Superseded by ISO/IEC 80000:2006-2009.
2. R. Dybkær, Standardisering af enhedsbetegnelser i klinisk-kemisk laboratoriearbejde, *Nord. Med.*, 1960, **63**, 26–31.
3. N. Radin, What is a standard?, *Clin. Chem.*, 1967, **13**, 55–76.
4. R. Dybkær and K. Jørgensen, IUPAC–IFCC (International Union of Pure and Applied Chemistry, and International Federation of Clinical Chemistry, Commission on Clinical Chemistry). *Quantities and Units in Clinical Chemistry, including Recommendation 1966*, Munksgaard, København, 1967.

5. ICSH; IFCC; WASP, Recommendations for the use of SI in clinical laboratory measurements, *Br. J. Haematol.*, 1972, **23**, 787–788. Also *Z. Klin. Chem.*, 1973, **11**, 93; *IFCC Newsletter* 1973, **08**(8). Es: Cinco resoluciones preparadas pos el Panel de Expertos en Unidades y Medidas (IFCC) y la Comison de Unidades y Medidas en Quimica Clinica. *Acta Bioquim. Clin. Latinoam.*, 1976, **10**(4), 362.

6. IUPAC; IFCC, 1972. *Quantities and Units in Clinical Chemistry*. Information Bulletin, Appendices on tentative nomenclature, symbols, units, and standards – Number 20.

7. IUPAC; IFCC, 1972. *List of Quantities in Clinical Chemistry*. Information Bulletin, Appendices on tentative nomenclature, symbols, units, and standards – Number 21.

8. R. Dybkaer, The special name "katal" for the SI derived unit, mole per second, when expressing catalytic activity, *Metrologia*, 2000, **37**, 671–676.

9. BIPM; IEC; IFCC; ISO; IUPAC; IUPAP; OIML, *International Vocabulary of Basic and General Terms in Metrology*, ISO, Geneva, 2nd edn, 1993.

10. BIPM; IEC; IFCC; ILAC; ISO; IUPAC; IUPAP; OIML, 2012, *International vocabulary of metrology – Basic and general concepts and associated terms VIM*, 3rd edn. JCGM 200: 2012. This 3rd edition is also published as ISO Guide 99 by ISO (ISO/IEC Guide 99-12: 2007). Replaces the 2nd edn, 1993. Available on the Web site of the BIPM: http/www. bipm.org. (acc. 2015-12).

11. D. A. Lowe, *A Guide to International Recommendations on Names and Symbols for Quantities and on Units of Measurement*, WHO, Geneva, 1975.

12. WHO, 1977. *The SI for the Health Professions, Prepared at the Request of the Thirtieth World Health Assembly*, WHO, Geneva, 1997.

13. N. Sheppard, H. A. Willis and J. C. Rigg, Names, Symbols, and Definitions of Quantities and Units in Optical Spectroscopy, *Pure Appl. Chem.*, 1985, **57**(1), 105–120. Also *J. Clin. Chem. Clin. Biochem.*, 1987, **25**(5), 327–333; *Labmedica*, 1987, **4**(2), 29–39; *Quím. Clín.*, 1988, **7**, 183–192.

14. J. C. Rigg, B. F. Visser and H. P. Lehmann, Nomenclature for derived quantities, *Pure Appl. Chem.*, 1991, **63**(9), 1307–1311. Supersedes, *Chem. Intern.*, 1985, **7**(3), 29–33.

15. R. Zender, Whims on VIM, *J. Int. Fed. Clin. Chem.*, 1992, **4**(3), 115–120.

16. R. Dybkær and K. Jørgensen, Measurement, value, and scale, *Scand. J. Clin. Lab. Invest.*, 1989, **49**(Suppl. 194), 69–76.

17. R. Dybkaer, Metrological problems and requirements in the life sciences, *J. Int. Fed. Clin. Chem.*, 1989, 104–107. *Scand. J. Clin. Lab. Invest.*, 1989, **49**(Suppl. 193), 112.

18. R. Dybkaer, *An Ontology on property for physical, chemical, and biological systems*. 2009. http://ontology.iupac.org. (acc. 2015-12).

19. B. F. Visser, *Logarithmic quantities and units* – O. *Siggaard-Andersen*, Private Press, Copenhagen, 1981, p. 23–30.

20. ISO 31-6:1992. *Quantities and units*. Part 6. *Light and related electromagnetic radiations*. Superseded by ISO 80000-7:2008. *Quantities and units*. Part 7. *Light*.

21. J. C. Rigg, S. S. Brown, R. Dybkær and H. Olesen, *Compendium of Terminology and Nomenclature of Properties in Clinical Laboratory Sciences. The Silver Book. Recommendations 1995*. IUPAC-IFCC, Blackwell, Oxford, 1995.

22. EN 12435:2006. *Healthcare informatics – Expression of results of measurements in health sciences*.

23. EN 1614:2006. *Healthcare informatics – Structure for nomenclature, classification, and coding of properties in clinical laboratory sciences*.

24. http://www.ifcc.org//ifcc-scientific-division/sd-committees/c-npu/ (acc. 2015-12).

25. http://www.labterm.dk/Enterprise%20Portal/NPU_download.aspx (acc. 2015-12).

SECTION 2
Definitions of Some Disciplines Applied in the Clinical Laboratory

2.1 Introduction

This section describes a set of terms and definitions of disciplines or specialities related to the knowledge and skills necessary for various types of clinical laboratory activities. The English language is used as *lingua franca* (not as the official language of any country); thus the terms for some disciplines or specialities are the English literal translation (with British spelling) of the terms used in different countries or regions. The names of disciplines or specialities directly related to the clinical laboratory, used in entries having a definition, are the names of specialities used in the English language version of the Official Journal of the European Union,[1] with two exceptions that will be explained later. Those terms should be considered as "neutral" or "stateless" ones, although each country or region has its own terms, including those countries having English as official or co-official language.

Few of the sciences and branches of study contributing to the work of the IUPAC Division of Chemistry and Human Health and to IFCC have internationally agreed definitions. Although the term **clinical chemistry** was incorporated in the name of the Federation in 1955, there are differences between countries in professional divisions, which perhaps partly explain differences in terms, *e.g.* **clinical pathology** in the USA, **biologie clinique** in Belgium and **clinical analyses** in Spain. There are also differences in functional divisions of staff between hospitals, *e.g.* a **clinical chemist** may also be manager of a blood bank. Some clinical chemists have primary training in chemistry or biochemistry; others in biology, medicine or pharmacy. As a group of scientific disciplines, their limits cannot be defined either in terms of place of work or object of study: for instance, veterinary applications are comparable to (human) medical applications. All these considerations make it difficult to define a single discipline for which these recommendations apply.

The following is an alphabetical listing of widely used terms with definitions for some of the disciplines, in which users of these recommendations may be working. Any definition is indicative and represents no recommendation. Instead of *discipline*, some definitions speak of *speciality.* The layout of the definitions is inspired by the ISO standard 10241-1,[2] and is generally systematically constructed.

Some experts prefer to distinguish the function of the disciplines in the service of medicine with the modifiers medicine, medicinal or medical. Others prefer the modifier clinical. There are national sensitivities around whether the disciplines are branches of medicine or health science disciplines. If the concepts are viewed as sciences, these different forms of description are interchangeable, but where they conflict with professional demarcation, these different forms of description are not interchangeable. Thus, in English-speaking countries, "clinical biology" is termed "clinical pathology" or "laboratory medicine" or "clinical laboratory sciences",

Compendium of Terminology and Nomenclature of Properties in Clinical Laboratory Sciences: Recommendations 2016
Edited by Georges Férard, René Dybkaer and Xavier Fuentes-Arderiu
© International Union of Pure and Applied Chemistry 2017
Published by the Royal Society of Chemistry, www.rsc.org

but the Union Européenne des Médecins Spécialistes does not agree with all these denominations and recommends "clinical biopathology".[1]

2.2   Vocabulary of Disciplines

2.2.1   **biological chemistry**
branch of clinical biology that is concerned with the *in vitro* examination of chemical and biological properties
NOTE 1: Detection of substances (or their derivatives) for diagnostic or therapeutic reasons and detection of poisons (or their derivatives) are properly included in the field.
NOTE 2: In this Compendium, "biological chemistry" means "human biological chemistry".
NOTE 3: There is a great diversity of terms among countries and regions. For example, various European Union countries have: *biological chemistry* (Luxembourg), *clinical chemistry* (Finland, the Netherlands and Sweden), *chemical pathology* (Ireland, Malta and United Kingdom), *clinical biochemistry* (Czech Republic, Denmark, Italy, Slovakia and Spain), *medical biochemistry* (Slovenia), and *medical and chemical diagnostic laboratory* (Austria).[1]

2.2.2   **biological haematology**
branch of clinical biology that is concerned with the *in vitro* examination of properties of blood cells and blood plasma
NOTE 1: There is a great diversity of terms among countries and regions. For example, various European Union countries have: *clinical blood type serology* (Denmark), *biological haematology* (Luxembourg) and *clinical haematology* (Portugal).[3]
NOTE 2: Biological haematology includes clinical cytohaematology.

2.2.3   **biotechnology**
branch of applied science that is concerned with the application of science and technology to living organisms, as well as parts, products and models thereof, to alter living or non-living materials for the production of knowledge, goods and services

2.2.4   **chemical pathology**
See biological chemistry (NOTE 3).

2.2.5   **clinical bacteriology**
branch of clinical microbiology that is concerned with the *in vitro* examination of properties related to bacteria in materials derived from the human body, or from another origin

2.2.6   **clinical biochemistry**
See biological chemistry (NOTE 3).

2.2.7   **clinical biology**
branch of health science that is concerned with the *in vitro* examination of properties in materials derived from the human body, or other origin, to provide information for the diagnosis, prevention and treatment of disease in, or the assessment of the health of human beings, by means of chemical or biological methods
NOTE 1: Although this definition might include all aspects of the speciality pathological anatomy except *post mortem* examinations, traditionally in most countries pathological anatomy has been considered independent of clinical biology.

NOTE 2: There is a great diversity of terms among countries and regions. For example, various European Union countries have: *clinical analyses* (Spain), *clinical biology* (Belgium and Luxembourg), *clinical pathology* (Italy and Portugal), *diagnostic laboratory* (Poland), *laboratory medicine* (Estonia and Lithuania), *medical biology* (Austria and France) and *medical laboratory diagnostics* (Hungary).[1]

NOTE 3: Because the specialities of the clinical laboratory are practised by graduates from many disciplines, the term *clinical laboratory sciences* is used or recommended in many countries, and may be regarded as a synonym for *clinical biology*.

2.2.8   clinical chemistry
See biological chemistry (NOTE 3).

2.2.9   clinical cytohaematology
branch of biological haematology that is concerned with the *in vitro* examination of properties of human blood cells, including their precursors and related particles

2.2.10   clinical cytology
branch of pathological anatomy that is concerned with the *in vitro* examination of human cell properties

2.2.11   clinical cytogenetics
branch of clinical genetics that is concerned with the *in vitro* examination of human chromosome properties

2.2.12   clinical genetics
branch of clinical biology that is concerned with the *in vitro* examination of human chromosome properties
NOTE: Clinical genetics encompasses clinical cytogenetics and clinical molecular genetics.

2.2.13   clinical genomics
branch of clinical molecular genetics that is concerned with the *in vitro* examination of the human genome properties

2.2.14   clinical haemostasiology
branch of clinical biology that is concerned with the *in vitro* examination of human haemostatic properties

2.2.15   clinical histology
branch of pathological anatomy that is concerned with the *in vitro* examination of human tissue properties

2.2.16   clinical immunology
branch of clinical biology that is concerned with the *in vitro* examination of immune system properties in materials derived from the human body
NOTE 1: The term *clinical immunology* given here is more appropriate than the term *immunology* used in the English language version of the Official Journal of the European Union.[1]
NOTE 2: The properties being examined may be related to antigens or antibodies.
NOTE 3: There is a great diversity of terms among countries and regions. For example, the various European Union countries have: *allergology and clinical immunology* (Czech Republic and Hungary), *clinical immunology* (Denmark, Ireland, Poland and Sweden), *clinical immunology and allergology* (Slovenia) and *immunology* (Austria, Cyprus, Latvia, Malta, Spain and United Kingdom).[1]

2.2.17   **clinical laboratory sciences**
See clinical biology (NOTE 3).

2.2.18   **clinical microbiology**
branch of clinical biology that is concerned with the *in vitro* examination of microbiological and parasitological properties of materials derived from the human body or healthcare materials
NOTE 1: The term *clinical microbiology* given here is more appropriate than the term *microbiology-bacteriology* used in the English language version of the Official Journal of the European Union.[1]
NOTE 2: There is a great diversity of terms among countries and regions. For example, various European Union countries have: *clinical bacteriology* (Sweden), *clinical microbiology* (Czech Republic, Denmark, Finland, Slovakia and Slovenia), *hygiene and microbiology* (Austria), *medical microbiology* (Hungary, the Netherlands, Poland), *medical microbiology and virology* (United Kingdom), *microbiology* (Cyprus, Greece, Ireland, Latvia, Luxembourg and Malta), *microbiology and infection epidemiology* (Germany), *microbiology and parasitology* (Spain) and *microbiology and virology* (Italy).[1]

2.2.19   **clinical molecular biology**
See clinical molecular genetics (NOTE 2).

2.2.20   **clinical molecular genetics**
branch of clinical genetics that is concerned with *in vitro* examination of properties of human genes, including their products
NOTE 1: Clinical molecular genetics encompasses clinical genomics, clinical pharmacogenomics and clinical proteomics.
NOTE 2: Clinical molecular genetics in some documents has been termed "clinical molecular biology".

2.2.21   **clinical mycology**
branch of clinical microbiology that is concerned with the *in vitro* examination of properties of fungi in materials derived from the human body or other origins

2.2.22   **clinical parasitology**
branch of clinical microbiology that is concerned with the *in vitro* examination of properties of parasites in materials derived from the human body or other origins

2.2.23   **clinical pathology**
See clinical biology (NOTE 3).

2.2.24   **clinical pharmacogenomics**
branch of clinical molecular genetics that is concerned with the *in vitro* examination of human gene properties to provide information for the pharmacological treatment of disease

2.2.25   **clinical proteomics**
branch of clinical molecular genetics that is concerned with the *in vitro* examination of properties of the human proteome

2.2.26   **clinical serology**
branch of clinical immunology that is concerned with the *in vitro* examination of properties related to antigens or antibodies in serum or plasma

2.2.27   **clinical toxicology**
branch of clinical biology that is concerned with the *in vitro* examination of toxic substance properties in materials derived from the human body or other origin

2.2.28 **clinical virology**
branch of clinical microbiology that is concerned with the *in vitro* examination of virus properties in materials derived from the human body or other origin

2.2.29 **health science**
branch of applied science that is concerned with the preservation and restoration of human and animal health

2.2.30 ***in vitro* diagnostics**
branch of applied science that is concerned with the development and production of examining systems for the *in vitro* examination of properties of materials derived from the human body or healthcare materials, to provide information for the assessment of health, or the prevention, diagnosis and treatment of disease

2.2.31 **laboratory medicine**
See clinical biology (NOTE 2).

2.2.32 **metrology**
science of measurement and its application [ref. 4, concept 2.2]
NOTE 1: Metrology is a branch of applied science.
NOTE 2: Metrology includes all theoretical and practical aspects of measurement, whatever the measurement uncertainty and field of application [ref. 4, concept 2.2, Note].

2.2.33 **molecular diagnostics**
branch of both biotechnology and clinical biology that is concerned with *in vitro* examination of properties related to the molecular structure and function of genes, including their products

2.2.34 **pathological anatomy**
branch of health science that is concerned with the *in vitro* examination of morphological properties of cells, tissues and organs, and the *post mortem* examination of the human body to provide information for the health assessment or the prevention, diagnosis and treatment of disease in human beings, by means of macroscopic inspection, microscopy, and chemical and biological techniques
NOTE 1: In most countries the examination of blood and bone marrow cells is included in biological haematology.
NOTE 2: Although pathological anatomy may be considered a branch of clinical biology, except for *post mortem* examinations, in most countries the discipline (and speciality) pathological anatomy has been traditionally considered independent of clinical biology.
NOTE 3: There is a great diversity of terms among countries and regions. For example, various European Union countries have: *clinical pathology* (Sweden), *histopathology* (Malta and United Kingdom), *morbid anatomy and histopathology* (Ireland), *pathology* (Austria, Estonia, Finland, Germany, Hungary, Latvia, Lithuania and the Netherlands), *pathological anatomy* (Belgium, Cyprus, Czech Republic, Greece, Italy, Luxembourg, Portugal, Slovakia and Spain), *pathological anatomy and cytology* (Denmark, France), *pathological anatomy and cytopathology* (Slovenia) and *pathomorphology* (Poland).[1]

2.2.35 **therapeutic drug monitoring**
branch of health science that is concerned with the *in vitro* measurement of drug concentrations in materials derived from the human body to monitor and guide the pharmacological treatment of disease
NOTE: Depending on the countries, therapeutic drug monitoring practitioners may do their work in the clinical laboratory or in the hospital pharmacy service.

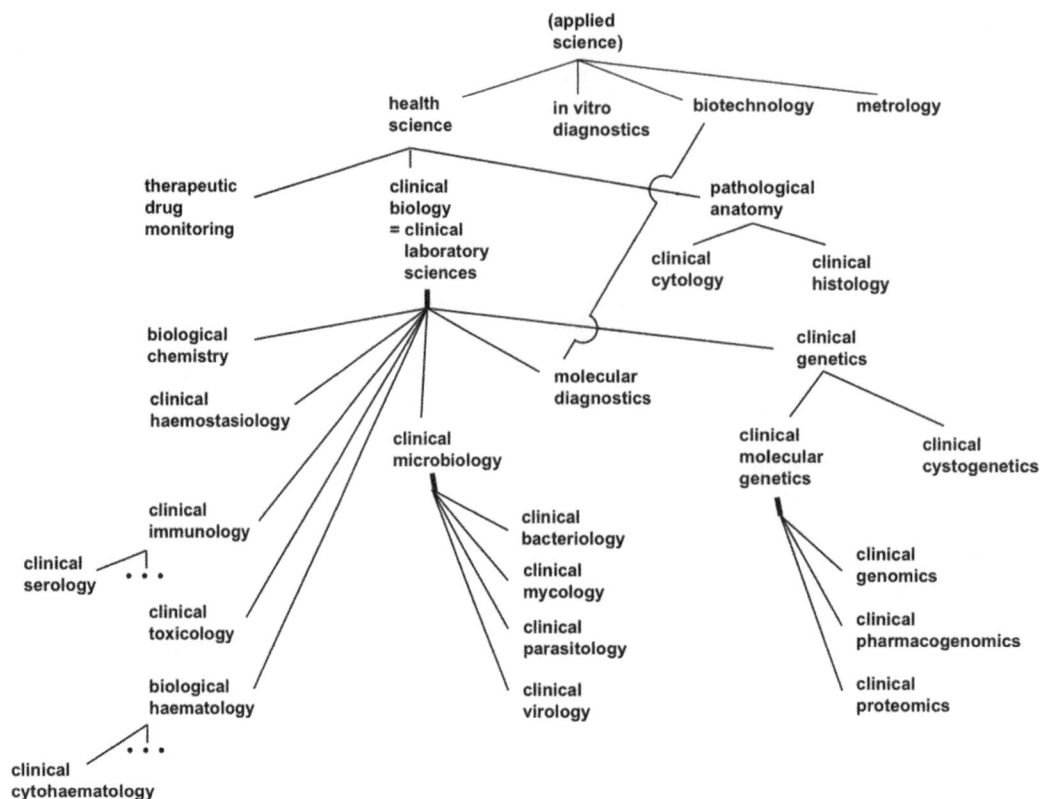

Figure 2.1 Generic concept diagram around 'clinical biology' = 'clinical laboratory sciences'. A short generic relation line to three bullets indicates that one or more specific concepts are possible.

2.3 Relations between Disciplines

The disciplines defined in subclause 2.2 are hierarchically related as shown in the generic concept diagram (Figure 2.1), constructed according to ISO 704.[5] To avoid an overly complicated diagram, not all the possible associative relations for metrology are indicated.

References

1. European Parliament, Council of European Union. *Council Directive 93/16/EEC of 5 April 1993* to facilitate the free movement of doctors and the mutual recognition of their diplomas, certificates and other evidence of formal qualifications. http://ec.europa.eu/internal_market/qualifications/specific-sectors_doctors_en.htm. (acc. 2015-12).
2. ISO 10241-1:2011. *Terminology entries in standards – Part 1: General requirements and examples of presentation.* Supersedes ISO 10241-1:2000.
3. http://www.labterm.dk/Enterprise%20Portal/NPU_download.aspx (acc. 2015-12).
4. BIPM; IEC; IFCC; ILAC; ISO; IUPAC; IUPAP; OIML, 2012. *International vocabulary of metrology – Basic and general concepts and associated terms VIM*, 3rd edn. JCGM 200: 2012. This 3rd edition is also published as ISO Guide 99 by ISO (ISO/IEC Guide 99-12: 2007). Replaces the 2nd edition, 1993. Available on the Web site of the BIPM: http//www.bipm.org. (acc. 2015-12).
5. ISO 704:2000. *Terminology work — Principles and methods.*

SECTION 3

Conventions and Instructions for Use

The primary purpose of this Silver Book is to be a guide towards a structured and uniform way of reporting on examinations from clinical laboratories, but it also provides the means to interpret older terms of properties and recalculate values expressed in older units.

For the purpose of this Compendium, **terminology** is taken as a set of designations belonging to one special language and **nomenclature** as terminology structured systematically according to pre-established naming rules (ISO 1087-1[1]). The recommendations of the International Vocabulary of Metrology from JCGM-200,[2] ISO/IEC 80000 series,[3] IUPAC Green Book[4] and Gold Book[5] have been taken into account.

3.1 Typographical Conventions

3.1.1 Spelling

For words of Greek origin, the Latinized form has been taken in preference to the simplified spelling, but the choice of spelling represents no part of the recommendations.

3.1.2 Hyphenation

The term **amount-of-substance** is here hyphenated in order to emphasize that it designates a single concept. The same convention applies for amount-of-energy, -of-light, -of-heat, -of-catalyst, and -of-enzyme as well as for kind-of-property (Sections 4 and 10), and kind-of-quantity.

3.2 Indication of Synonyms and Choice between them

Many of the concepts used have several terms, sometimes *full synonyms* with complete or almost complete overlap of meaning, sometimes *partial synonyms* with only partial overlap of meaning. In text, synonyms are indicated by a round bracket introduced by 'or'. Synonyms considered equally acceptable are printed in bold type, less acceptable synonyms in italic type.

Examples

☐ **photon** (or **quantum**) quantum being equally acceptable
☐ **massic volume** (or *specific volume*), massic volume being preferred (§9.15.1)

The choice of preferred synonym depends on several factors, notably disciplinary practice and language. In this Compendium, the terminological principles given in §5.4 are

Compendium of Terminology and Nomenclature of Properties in Clinical Laboratory Sciences: Recommendations 2016
Edited by Georges Férard, René Dybkaer and Xavier Fuentes-Arderiu
© International Union of Pure and Applied Chemistry 2017
Published by the Royal Society of Chemistry, www.rsc.org

followed. This means that the first choice of term is sometimes found in second place in ISO/IEC 80000, *e.g.* as in the second example above.

In the lists of kinds-of-quantity (Sections 8 and 9), the heading of each section is a systematic term, usually constructed according to principles described in §5.4. No preference is expressed between the systematic term in that heading and other current terms indicated as full synonyms in bold type. Lesser degrees of synonymy are indicated as *partial synonymy.* If so categorized, the term could be attached to several kinds-of-quantity and so is discouraged here, even if recommended by some authorities. Among the full synonyms, no preference is normally expressed except where terms are unduly long or where they include certain words without a clear meaning. In contrast to other full synonyms, they are then not highlighted in bold-face letters.

In Sections 8 and 9, usages that are known to be discouraged are marked by a dagger (†). *Example*

☐ † A.U. absorbance unit

Some of the more specialized usages of current terminology are listed as examples rather than as partial synonyms. They are then given in the format recommended for specification of properties in the clinical laboratory (§6.4) followed by current synonym(s) and symbol.
Examples

☐ Air—Water vapour; relative mass concentration; **relative humidity**; h (§8.4.4)
☐ System—Carbon dioxide; substance concentration; $c(CO_2)$. *Partial synonyms*: CO_2 concentration; carbon dioxide concentration (§9.88.1) (which could alternatively mean mass concentration, §9.14.2).

3.3 Definitions

In taking definitions from various sources, the wording has been unified in style in line with the conventions of the International Standard ISO 10241-1.[6] Where a source definition uses, for instance, the form "The ratio of X to Y", this has been amended in this Compendium and completed in line with the standard to 'X of a component (or system) divided by the Y of the component (or system)'. Where several definitions of a concept are given, (1), (2), ... represent different definitions of what is apparently a single concept and 1, 2, ... represent apparently variant but related concepts under a single term.

For some of the more specialized kinds-of-property only symbolic definitions are given.

3.4 Indication of Recommendations and of Advice

The auxiliary verb ***should*** is used to indicate preferences in this book, not binding, arising, for instance from logic or coherence; ***shall*** is used only where IUPAC,[4] IFCC[7] or higher authorities such as CGPM,[8] ISO and IEC[9] have made recommendations.

3.5 Order of Entries in Sections 8 and 9

The entries of Sections 8 and 9 are arranged in the order of the dimensions of the base kinds-of-quantity given in Table 5.1. The order of entries for electrical kinds-of-quantity is based on relationship to the extensive kind-of-quantity electrical charge and its unit coulomb (§9.50U), not electrical current and the ampere (§9.52U). Likewise photometric kinds-of-quantity are classed on the basis of the extensive kind-of-quantity amount-of-light with the SI unit lumen second (§9.79U) and not on luminous intensity with the base SI unit candela (§9.81U). For kinds-of-quantity of dimension one, the system is further described at the beginning of Section 8.

3.6 Order of Information about Kinds-of-quantity in Entries in Sections 8 and 9

The elements of information are tagged, as appropriate, by numbers in angle brackets.

1 entry number (not tagged)
⟨2⟩ systematic terms (in bold type) followed by generalized specifications in normal type
⟨3⟩ synonym(s) (preferred ones in bold type)
⟨4⟩ abbreviation (§6.7.2)
⟨5⟩ preferred symbol(s)
⟨6⟩ definition in words
⟨7⟩ definition by equation
⟨8⟩ deprecated terms and symbols
⟨9⟩ note(s)
⟨10⟩ example(s) of dedicated kinds-of-quantity

3.7 Preference between Symbols for a Kind-of-quantity

Preference between symbols is taken in line with majority choices of authorities, especially the International Standard ISO/IEC 80000[3] and the IUPAC Green Book.[4] A few symbols are discouraged, in line with preference of international authorities, and are then listed under "deprecated terms and symbols". Symbols for kinds-of-quantity are, however, a matter of convenience and others may be chosen as long as meanings are given.

In Sections 8 and 9, symbols are listed with modifiers (§5.5.1) corresponding to some of the specifications attached to the term. In practical equations, these modifiers may not be necessary or may need to be replaced by particular specifications (§6.3).

Independently of the above preferences, IUPAC and IFCC recommend **abbreviations** of kinds-of-quantity rather than symbols for terms in clinical laboratory reports (§6.7.2 and Table 6.2).

3.8 Lists of Information about Units

At the end of each group of kinds-of-quantity in Sections 8 and 9, information about units is given under numbered tags in angle brackets:

⟨12⟩ term of SI (coherent) unit, in some instances distinguished for different kinds-of-quantity, in accordance with preference of ISO/IEC 80000 series[3]
⟨16⟩ definitions of base SI unit(s) according to the SI Brochure[8]
⟨19⟩ notes on use and usage of units

For convenience a reference page 'Structure of Numbered Entries in Sections 8 and 9' listing this information can be found on page xxiii of this new edition of the Silver Book.

A simple preference is expressed between recommended and other unit symbols. Criteria for a preference are given in Section 5: preference for SI and for certain ways of expressing multiples of SI units, including avoidance of units with prefixes to more than one component unit symbol; the preference in several clinical laboratory sciences of the litre for expression of concentrations. The status of units listed as 'other' varies in recommendations of the BIPM brochure and in regional (*e.g.* European Union) and national legislation (*e.g.* United States or Canada). It is therefore not the intention that the tables show which units are legal, but rather which are more advisable and less advisable. However, when an authoritative body has given a clear position, this is adopted in this Compendium. Because of the complexities of where the plural endings of compound units are placed in English, for instance those with the word *per*, the plural of such units is indicated, usually in parentheses. For the more common kinds-of-quantity such as length, mass and volume, some traditional units are included to allow conversion of older data.

There are no rules for the order of component units in the terms and symbols of compound units. For consistency, positive powers are usually given before negative powers; within the groups of positive and negative powers, the order of SI base units is followed.

3.9 Inconsistencies and Points for Debate Remain

In the general guidelines (Sections 4–7) and in definitions of kinds-of-quantity (Sections 8 and 9), recommendations of about 50 years have been updated and inconsistencies have been resolved as far as possible. In Section 10 examples are given regarding kinds-of-property without dimensions of the ISQ. This large domain is currently being structured[10] and recommendations will eventually be published. Even with the benefit of international recommendations, symbols can be ambiguous. However, ambiguities can be reduced if symbols are distinguished into the categories and contexts defined in Sections 4 and 5. A list of symbols for kinds-of-quantity used in clinical laboratory sciences is provided at the end of this Compendium.

Example

☐ The distinction between Mg for megagrams (1000 kg) and Mg for magnesium should be clear from the context.

Systematic terms used in Sections 8 and 9 are in line with recommendations of IUPAC-CQUCC and IFCC-CQU,[11] and ISO/IEC 80000.[3] Others are used here for the first time, and might be improved.

References

1. ISO 1087-1:2000. *Terminology work — Vocabulary* — Part 1: *Theory and application.*
2. BIPM; IEC; IFCC; ILAC; ISO; IUPAC; IUPAP; OIML, 2012. *International vocabulary of metrology – Basic and general concepts and associated terms VIM*, 3rd edn. JCGM 200: 2012. This 3rd edition is also published as ISO Guide 99 by ISO (ISO/IEC Guide 99-12: 2007). Replaces the 2nd edition, 1993. Available on the Web site of the BIPM: http/www.bipm.org. (acc. 2015-12).
3. ISO/IEC 80000:2006-2009. *Quantities and units* (14 parts). Supersedes ISO 31:1992.
4. I. Mills, T. Cvitaš, K. Homann, N. Kallay and K. Kuchitsu, *Quantities, Units and Symbols in Physical Chemistry — The IUPAC Green Book*, RSC Publishing, Cambridge, 3rd edn, 2007.
5. *IUPAC Compendium of Chemical Terminology — the Gold Book*, RSC Publishing Cambridge, 2005-2014. http://goldbook.iupac.org/.
6. ISO 10241-1:2011. *Terminology entries in standards* – Part 1: *General requirements and examples of presentation.* Supersedes ISO 10241-1:2000.
7. J. C. Rigg, S. S. Brown, R. Dybkær and H. Olesen *Compendium of Terminology and Nomenclature of Properties in Clinical Laboratory Sciences. The Silver Book. Recommendations 1995.* IUPAC-IFCC, Blackwell, Oxford, 1995.
8. *Le Système International d'Unités (SI), The International System of Units (SI)*, 8th edn. (Bureau International des Poids et Mesures, Sèvres, France, 2006); it is the official French text followed by an English translation; commonly called the BIPM SI Brochure. Available on the Web site of the BIPM: www.bipm.org/en/si/ (acc. 2015-12).
9. ISO/IEC:2009. *Information technology – Security techniques – Information security management systems – Overview and vocabulary.*
10. G. Nordin, R. Dybkaer, U. Forsum, X. Fuentes-Arderiu, G. Schadow and F. Pontet, An outline for a vocabulary of nominal properties and examinations – Basic and general concepts and associated terms, *Clin. Chem. Lab. Med.*, 2010, **48**(11), 1553–1566.
11. J. C. Rigg, B. F. Visser and H. P. Lehmann, Nomenclature for derived quantities, *Pure Appl. Chem.*, 1991, **63**(9), 1307–1311. Supersedes, *Chem. Intern.*, 1985, **7**(3), 29–33.

Fundamental Concepts in Communication of Clinical Laboratory Information

4.1 Communication of Information

4.1.1   The main task of the clinical laboratory scientist in the health system is to obtain and communicate information on the chemical, biochemical, physiological, pathological and sometimes physical properties of people. Such information is needed for prevention, diagnosis, prognosis and therapy of diseases, and to learn about disease at the molecular level. It is essential, therefore, that the properties examined in the clinical laboratory are those needed by the clinician and that the results communicated from the clinical laboratory to the clinician are correctly understood.

4.1.2   To avoid misunderstandings, **a set of rules is needed for transmission of information**. Until the joint Recommendations approved in 1966,[1] there was no accepted usage in the discipline of clinical laboratory sciences internationally, nationally or locally, causing inconvenience and waste of time for staff, and sometimes danger to the patient.

4.1.3   Besides diverse and quaint usages within the discipline, there were communication barriers between clinical laboratory sciences and relevant branches of chemistry, biology and physics. Since the 1960s, the community of clinical laboratory scientists has been increasingly recognized as part of the scientific community and has played its part in standardizing language in many areas of applied science, for instance VIM.[2] Since the 1990s, communications have been established and consensus obtained with several organizations representing different clinical laboratory sciences.[3–13]

4.1.4   Most clinical laboratories provide other health services with information about a large number of different **types of property** of the human body. The laboratorians strive to present measurement results for **quantities**, *i.e.* those properties that have a magnitude, mostly expressed numerically. Still, many other important **properties** of the type of **nominal properties** (see Section 10), *e.g.* related to species of microbes or genetic structure, are inherently devoid of magnitude, but also require a place in a comprehensive nomenclature. In that respect the third edition of the VIM,[2] in accordance with the concept 'metrology', is insufficient.

Compendium of Terminology and Nomenclature of Properties in Clinical Laboratory Sciences:
Recommendations 2016
Edited by Georges Férard, René Dybkaer and Xavier Fuentes-Arderiu
© International Union of Pure and Applied Chemistry 2017
Published by the Royal Society of Chemistry, www.rsc.org

4.1.5   As a logical basis for the structure of information about all types of property examined or measured in the clinical laboratory, the recommendations of 1966[1] took concepts from the disciplines of systems analysis and metrology: **system** (§4.2), **component** (§4.3) and **kind-of-quantity** (§5.1). In most examples, these elements of information are given in the NPU format recommended for the requesting and reporting of information from the clinical laboratory (§6.3):

System—Component; kind-of-property

4.1.6   The nature of components commonly encountered in the clinical laboratory has consequences for the **choice of the kind-of-quantity** (Section 7) that is most informative.

4.2   System

4.2.1   **Any examination**, including a physical or chemical measurement, is related to a **property of a specific system**, which in human biology is a **composite object** such as an entire body or a specified part of a body. A **system** can be defined as a demarcated part or phenomenon of the perceivable or conceivable universe, material or immaterial, that may be regarded as a set of entities, together with a set of relations or processes between these entities [modified ref. 27]. A system may be examined at a given clock and calendar time (§6.8). In analytical and preparative chemistry, the system may take the form of a **mixture** (§4.3.13), which is an alternative term used in some definitions, for instance of ISO/IEC 80000-9.[14]

Examples
□ Person, Kidney, Blood, Leukocytes, Urine, Beta cells of the pancreas, Exhaled air

4.2.2   If matter or energy can be introduced into or removed from a system it is an **open system**. If not, it is a **closed system**. All biological systems are open. However, for a certain property, the system may be regarded as closed.

Examples
□ Intestine(Patient)—Xylose absorption; amount-of-substance(time interval), where the intestine is considered as an open system.
□ Patient—Body; mass, where the patient can be considered as a closed system in an appropriately small time interval.

4.2.3   In macroscopic or microscopic terms, a system may appear homogeneous or heterogeneous. A **heterogeneous system** contains more than one **phase**, each phase being in the solid, liquid or gas state. Any phase may be specified as a system in itself. Alternatively, it may be convenient to consider a phase as a **subsystem** or **component** of the larger system.

Example
□ Blood is heterogeneous, and has plasma and cells as phases.

4.2.4   In clinical laboratory sciences, the system may be an entire **living organism**, usually a human being, commonly termed a **patient**, defined by date of birth, name and address, or by a personal identification number (§6.3.2). The system is also defined by **calendar date** and **time of day** (§6.8). Sometimes the system may be **part of an organism**, for instance

– an organ, *e.g.* pancreas
– a tissue, *e.g.* islets of pancreas
– a collection of cells of a certain type, *e.g.* beta cells
– a chemical entity, *e.g.* insulin

The system studied in the clinical laboratory may be prepared from the biological material, *e.g.* blood plasma is prepared from blood. The specification of the system (§6.5) may be brief, but must be sufficient for the purpose.

4.2.5	It is commonly unnecessary or impractical to use the entire biological system for an examination. Instead it is convenient to take a **sample**,[15] a representative portion of the specified system at a stated time. If the system sampled is continuously changing, the sample is sometimes termed a "**specimen**". The sample may be considered a system in itself. It cannot always be truly representative, for instance of blood, but at least it can be taken by a standardized procedure, for instance blood from a peripheral vein with minimum stasis. A sample may be further specified as **primary sample, laboratory sample** or **analytical sample**.[15] A primary sample is the sample directly drawn from the patient. A laboratory sample can be a sample used for several examinations in a common part of the laboratory. An analytical sample can be the sample used for a specific instrument or examination.
NOTE: In this Compendium, the term sample is not used in the sense of a random part of a human population.

4.2.6	Examinations may be made on the analytical sample as such or on smaller **analytical portions**, which must be representative of the analytical sample. The word "**aliquot**" is sometimes used when the analytical portions represent integral fractional amounts of the analytical sample.

4.2.7	The system under investigation may be a **population** of organisms and their environment. The population or a sample, *i.e.* a **representative group** of individuals from it, may be investigated. In environmental studies, for instance, the atmosphere in a building may be treated as a system.

4.2.8	The system may form the operative part of a **supersystem** or a **subsystem** may form the operative part of the system. Specifications about the supersystem or subsystem are conventionally placed (without space) in parenthesis after the name of the system in reports from the clinical laboratory:
System(Specifications about the system)—Component
Example
☐ If respiration is being studied, it may be considered as a function of the lungs or the whole organism, either of which may be taken as system:
Patient(Lungs)—Respiration, 'Lungs' being the subsystem or
Lungs(Patient)—Respiration, 'Patient' being the supersystem
Note that subscripts may be added to specify a system. The modifiers 1, 2, ..., n may be added to represent a series of systems. The modifier 0 (zero) is reserved for a reference system. An unspecified member of a series of systems may be represented by the subscript symbol for a running number, i, j or k, for instance in summation. For examples, see Sections 8 and 9.

4.3   Component and Entity

The concept of 'system' can be partitively divided into 'matrix' and 'component' parts of the system, where "component" is defined as "part of a system".

4.3.1	If a system, designated system 1, contains a component, designated component B, the relationship of system and component can be represented in terms of set theory as
component B ⊆ system 1
The component may consist of a set of **elements**, also designated B, with a common characteristic:
element B ∈ component B
In chemistry, the term *element* from set theory is rarely used to avoid confusion with the term **chemical element** (§4.3.4). Instead the term **entity** is used (§4.3.3).

4.3.2	A material system consists of **matter**, with one or more **chemical substances**, which may react with one another to form new substances. The purpose of chemistry is to describe substances and their reactions. Any definable substance or group of substances in a system may be considered as a **chemical component**.

Examples
- ☐ Haemoglobin in an erythrocyte
- ☐ Helium in air
- ☐ Water in blood
- ☐ Sodium ion in sweat
- ☐ Cholesterol plus cholesterol-containing substances in blood plasma
- ☐ Bacteria in urine
- ☐ Coagulation(process) in plasma

Usage of the term **chemical component** overlaps with that of **chemical species**, **chemical substance** and, in analytical chemistry, with **analyte**. In a similar way to chemical components, physical structures may be considered as **physical components** of a system (§4.3.11).

4.3.3   A principle of chemistry is that matter is built up of extremely small particles that may dissociate into still smaller particles or combine into larger particles. Any of these things may, in a more general term, be termed **entities**. A chemist may need to specify a component by **type of entity** (*e.g.* atoms, molecules, ions, an active group), by term or formula indicating the structure, by a class distinguished by the interval of a property (*e.g.* size, energy), or by a structural feature of the molecule (*e.g.* a double bond). The main types of **chemical entity** of interest in clinical laboratory sciences are described in §4.3.4–10. This view of matter as consisting of entities, sometimes also termed **elementary units** or **formula units,** is the foundation for the idea of amount-of-substance (§9.83.1).

4.3.4   The type of chemical entity may alternatively be a molecule, ion, **radical** or atom, which are the fundamental entities of a composed chemical component and should be specified by a complete term, in accordance with the following rules:

    (1) IUPAC for recommended terms in **inorganic chemistry**[16]
       Example
       ☐ Calcium ion(Ca^{2+}), not Ca or calcium

    (2) IUPAC for recommended terms in **organic chemistry**[17]
       Example
       ☐ Ethanol (rather than alcohol), or ethyl alcohol

    (3) IUBMB for **enzymes,**[18] designated either by systematic terms or by their recommended trivial terms, together with their enzyme codes (EC)
       Example
       ☐ Alkaline phosphatase (recommended trivial term), corresponding to orthophosphoric-monoester phosphohydrolase(alkaline optimum) (systematic term) (EC 3.1.3.1)

    (4) UPAC/IUB-CBN and IUB-NC for simplified terms of **common chemical entities** in biochemistry[19]
       Example
       ☐ Glucose (recommended trivial term) corresponding to *gluco*-hexose (systematic term). In the clinical laboratory, the configuration D is implicit because of the biological context.

    (5) Detailed steric information is necessary if several chemical entities have the same simplified formula[20]
       Example
       ☐ 11β-Hydroxyandrosterone

    (6) The chemical entity chosen for complex molecules may be represented by an appended symbol.

Examples
- ☐ Haemoglobin(Fe) for monomeric haemoglobin, a quarter of the tetrameric molecule
- ☐ Haemoglobin(Fe_4) for tetrameric haemoglobin

(7) WHO for trivial names (International Nonproprietary Names, INN) of some **biologically active chemical substances**[21] and of some pharmaceutical substances.[21,22]

Example
- ☐ Paracetamol (INN) corresponding to *N*-(4-hydroxyphenyl)acetamide (systematic term)

(8) ISO for trivial terms of **pesticides**,[23] formed by rules similar to those of INN

(9) **Chemical entities** may also be defined in physical terms, for instance by electrophoretic mobility (§9.55.1).

4.3.5 The type of entity may be defined as a **structural feature (active group or functional group)** common to several chemical entities.

Examples
- ☐ Triglycerate
- ☐ Monocarboxylate(–COO–)
- ☐ Monocarboxylate ion(–COO$^-$)
- ☐ Methyl ketone
- ☐ Fatty acid(carboxyl)
- ☐ Open chain of 6 carbon atoms(C_6)

4.3.6 The entity may be a **functional component (or a function group)** of a chemical compound or group of compounds, characterized by a common reactive property.

Examples
- ☐ Phosphatase
- ☐ Human immunodeficiency virus 1 antibody

4.3.7 If the type of chemical entity is a **molecule,** including **macromolecules** up to macroscopic size, a basic pattern may be repeated, with a formula in which *n* represents a large integer.

Examples
- ☐ Amylose, often symbolized $(C_6H_{10}O_5)_n$
- ☐ Poly(ethylene glycol) $(-CH_2-CH_2O-)_n$

NOTE: The number of repetitions, *n*, in a macromolecule may depend on temperature and pressure.

Macromolecules are usually described by their average molar mass (§9.92.1), often expressed as a relative molar mass (or "molecular weight") (§8.4.5).

Example
- ☐ System—DNA fragment; molar mass $= 1.2$ mg mol^{-1}

4.3.8 If the type of entity is an **ion,** it can be termed and symbolized in the same way as molecules with specification additionally of **state of ionization,** *i.e.* charge number, *z*, including the sign on the charge number, positive or negative. In chemical symbols, the state of ionization is indicated as a right superscript. Terms can be generated by adding the word **ion** to the term for the corresponding molecule or group.

Examples
- ☐ Na^+, Sodium ion
- ☐ Ca^{2+}, Calcium ion
- ☐ Cl^-, Chloride ion
- ☐ $SO_4{}^{2-}$, Sulfate ion

A component may comprise a set of entities. For example, the term 'calcium' is incomplete and should be defined by its entity as
Examples
- □ Calcium ion(0.5 Ca^{2+}), alternatively Calcium ion($z=1$), representing the **charge entity**, which is half of the ionic form
- □ Calcium ion(Ca^{2+}), alternatively Calcium ion($z=2$), representing the **ionic entity**
- □ Calcium($z=0$), representing the metallic element, which is not a component in a biological system

4.3.9 The oxidation number of a radical may be indicated with roman numerals or zero in parenthesis after the term or superscript after the symbol.
Examples
- □ Calcium(II) or Ca^{II} with an oxidation number +II, representing calcium either in ionic form or in chelates, for which ionic state is difficult to define.
- □ Calcium(0.5 Ca^{II})
- □ The combination of ionized, chelated and other forms of calcium(ɪɪ), but not the metal (see §4.3.8), sometimes termed "total calcium"

4.3.10 **Ionized and non-ionized forms** may coexist in the body but there is no general rule for terming such components.
The following practice is therefore suggested:
(1) The maximally ionized form is termed, omitting the word *ion*.
Examples
- □ Ammonium, comprising ammonium ion and ammonia
- □ Creatininium, comprising creatininium ion and creatinine
- □ Ascorbate, comprising ascorbate ion and ascorbic acid
- □ Carbonate, comprising carbonate ion, hydrogen carbonate ion and carbonic acid, but not carbon dioxide
- □ By contrast, ammonia, ascorbic acid, and carbon dioxide, each comprises a single chemical compound
(2) Trivial names for organic amphiionic components represent the totality of the amphiionic form (with charge or net charge zero), acid form and base form.
Example
- □ Hydroxyproline means hydroxyprolinium cation, hydroxyproline (uncharged) and hydroxyprolinate anion.
(3) Mixtures of a chemical substance and its derivatives are sometimes indicated by the plural of the term for the parent substance.
Example
- □ Cholesterols means a mixture of cholesterol and cholesterol esters.

4.3.11 Apart from the chemical entities, a material system can also contain one or more cytological, immunological, microbiological or parasitological entities.
Examples
- □ Blood—Erythrocytes
- □ Lymphocytes—CD3 antigen
- □ Urine—Bacteria
- □ Faeces—Parasite

The main cytological entities studied in the clinical laboratory are the **blood cells**, which are termed in line with the International Anatomical Committee's *Nomina anatomica, histologica, cytologica*[24] and in veterinary applications in line with the International Committee on Veterinary Gross Anatomical Nomenclature (ICVGAN).[25]

Bacteria, viruses, fungi, plants and **animals including parasites** should be designated by their **taxonomic term**: terms for genera, species and subspecies are printed in italic; terms for orders and families and for strains or races are printed in roman script (§6.6.4, §6.6.5, §6.6.6, §6.6.7.1 and §6.6.7.2).

4.3.12 The term for the **principle, method** or **procedure** of examination cannot normally replace the term for the component but forms a specification of the kind-of-property (§6.7.3).
Example
☐ Plasma(Blood)—Protein; mass concentration(colorimetry)

4.3.13 **Mixtures** (or **formulations**) of substances are commonly designated by an eponym, often after the originator, *e.g.* Fehling's solution. Such names are often ambiguous because the originator formulated various mixtures or because of variation from the original description. Note that the term "*mixture*" is also used in the sense of 'system' as defined in §4.2.1.

4.3.14 Some subscripts may be added to specify a component. The modifier A is reserved for the major component, *e.g.* a carrier gas or a solvent; the modifier E is used for an enzymic component and R for a reference component. An unspecified member of a series of components may also be represented by a subscript symbol for running number. For examples see Sections 8 and 9.

4.4 Process

4.4.1 In a clinical laboratory, an examined property is often intended to provide information about a biological process. Sometimes a process is part of the definition of the quantity.
Example
☐ Blood—Coagulation; time
According to the nature of the process and the examination procedure, the process may be indicated by the term or specification of the system, the component or the kind-of-property.

4.4.2 Transported matter entering or leaving an organism may be considered
– as a **system** in itself
Examples
☐ food considered for content of a nutrient:
Food—Sucrose; substance content
☐ urine considered for presence of glucose:
Urine—Glucose; substance concentration
– as a **component** of the organism
Examples
☐ food intake of vitamin A by the organism:
Patient—Retinol ingestion; substance rate (§9.94.1)
☐ rate of excretion of urea (or carbamide) by the organism:
Patient(Urine)—Urea excretion; substance rate (§9.94.1)
☐ Lungs—Water evaporation; mass rate (§9.26.1)
☐ Glomeruli(Kidneys)—Fluid filtered; volume rate (§9.24.1)
☐ Patient(Urine)—Phosphate(inorganic) excretion; substance rate (§9.94.1)
☐ Pancreas(Patient)—α-Amylase production; catalytic activity rate(30 to 150 min after meal; procedure) $= 18$ μkat s^{-1} $= 18$ μmol s^{-2} (§9.91.1)
Similarly matter transported into an organ may be considered as a system in itself or as a component of the organ.
Example
☐ Blood(Liver)—Perfusion; volume rate
☐ Liver—Blood(perfusion); volume rate

4.4.3   A component may need to be specified by the process in which it participates. No preference is given between adjectival form (*e.g.* excreted, filtered, produced) or nominal form (*e.g.* excretion, filtration, production) (examples under §4.4.2).

4.4.4   Chemical, physical and physiological processes may, themselves, be treated as components.
  Examples
  ☐ Blood—Coagulation
  ☐ Patient—Capillary bleeding
  ☐ Solution(Cuvette)—Radiation(wavelength; direction) absorbed

4.4.5   **Chemical processes** involve **conversion** (or **reaction**), in which a component may be formed or consumed. By convention, these specifications have become part of the terms of quantity used in chemical kinetics.
  Examples
  ☐ Extent of reaction of component B in system 1 is the increase in amount-of-substance of component B in a defined reaction in that system. It can alternatively be specified as change in amount-of-substance of component ν^{-1}B (§9.83.1, Note 7).
  ☐ Rate of conversion of component B in system 1 is the substance rate of component ν^{-1}B formed (§9.94.2).
  ☐ Rate of reaction of component B in system 1, as currently defined, is substance concentration rate of ν^{-1}B formed (§9.97.1).
  Conventions for equations of chemical reactions and electrochemical reactions are given in the IUPAC Green Book.[26] For chemical reactions, it recommends, for instance,

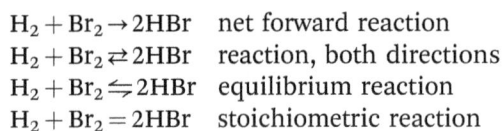

$H_2 + Br_2 \rightarrow 2HBr$ net forward reaction
$H_2 + Br_2 \rightleftarrows 2HBr$ reaction, both directions
$H_2 + Br_2 \leftrightharpoons 2HBr$ equilibrium reaction
$H_2 + Br_2 = 2HBr$ stoichiometric reaction

4.4.6   Terms commonly used to describe quantitative aspects of **physiological processes** in biological systems have no clear meaning without further explanation.
  Examples
  ☐ Rate of excretion is variously used for a volume, mass or amount-of-substance divided by time, perhaps also divided by volume or mass of the organism.
  ☐ Rate of growth is variously used for a number, length, volume or mass divided by time. Ways of specifying times and time intervals in such processes are given in §6.8.

4.5   **Property**: state- or process-descriptive feature of a system including any pertinent components [ref 27 modified]
  Example
  ☐ In the IFCC-IUPAC format (NPU format, see §6.4.1): Blood—Haemoglobin(Fe); amount-of-substance concentration with spatiotemporal specifications to the system 'Blood'.
  NOTE 1: A system is described by its properties.
  NOTE 2: The terms "attribute", "characteristic", "quality" and "trait" are sometimes used as synonyms of "property", but not in this Compendium.

4.6   **Kind-of-property**: aspect common to mutually comparable properties
  Examples
  ☐ amount-of-substance, concentration (or substance concentration), mass fraction, taxon, colour, catalytic-activity concentration, and entitic amount-of-substance

NOTE: The term is hyphenated to demonstrate that the phrase is a single term. The definition is analogous to that of 'kind of quantity' in the VIM [ref. 2, concept 1.2].

4.7 **Dedicated kind-of-property**: kind-of-property with given sort of system and any pertinent sorts of component[27]

Examples

☐ Patient—Urine; colour

☐ Plasma(Blood)—Sodium ion; amount-of-substance concentration

NOTE: The items necessary to indicate an individual property, relating to a property value, are the spatiotemporal coordinates. Omission of those leaves the dedicated kind-of-property concept.

4.8 **Type of property**

The superordinate concept 'property' can be generically divided in many ways according to purpose, but the sometimes used division into 'qualitative property', 'semiquantitative property' and 'quantitative property' should be discouraged because their perceived delineations vary. Another possibility (retained in this Compendium) is to distinguish different types of property according to the algebraic characteristics of the types of properties (Table 4.1) and their values will define the subordinate specific concepts – sometimes termed types of property – as follows.[27–29]

The algebraic restrictions naturally influence the types of statistics that can be applied to the property values.

4.8.1 **Nominal property**: property of a phenomenon, body or substance where the property has no magnitude [ref. 2, concept 1.30]

NOTE 1: As negative definitions are deprecated in terminology work, another definition has been proposed: property defined by an examination procedure that can be compared for identity with another property of the same kind-of-property, but has no magnitude (modified from ref. 27)

NOTE 2: The term "qualitative property" has also been used, but this is ambiguous because 'ordinal property' is often included under that term.

Examples

☐ Erythrocytes(Blood)—Erythrocyte antigen({ABO; Rh D}) = O+

☐ Blood—Plasma; colour = red

☐ Expectorate—Bacterium; taxon = *Mycobacterium tuberculosis*

☐ Urine—Neuroleptic drug; taxon(procedure) = chlorpromazine

☐ DNA(Leukocytes)—FRDA gene; sequence variation = (GAA)n expansion

4.8.2 **Ordinal quantity**: quantity, defined by a conventional measurement procedure for which a total ordering relation can be established, according to magnitude, with other quantities of the same kind, but for which no algebraic operations among those quantities exist [ref. 2, concept 1.26]

Table 4.1 Classification of types of property with their algebraic delimiting characteristics.

Type of property	Possible comparison[a]	Example of kind-of-property
nominal property	$a = b$; $a \neq b$	Blood group
ordinal property	$a < b$; $a = b$; $a > b$	Subjective temperature
linear differential property	$\lvert a - b \rvert = \lvert c - d \rvert$ $\lvert a - b \rvert \neq \lvert c - d \rvert$	Celsius temperature
logarithmic differential property	as above	pH
rational property	$a = nb$; $a \neq nb$	Thermodynamic temperature

[a]The symbol a refers to a property; b, c and d refer to properties in other systems, at other times, and under other conditions.

NOTE 1: **Ordinal property**, ordinal quantity: property defined by an examination procedure, having a magnitude that can be stated only to be lesser than, equal to or greater than another quantity of the same kind-of-property.[27] An ordinal scale serves for the comparison of examined values regarding a kind-of-property that can be ordered by magnitude.

NOTE 2: Ordinal properties can have alphabetical or numerical codes that form an ordinal scale of magnitude.

Examples
- ☐ Urine—Nitrite; arbitrary concentration({0, 1}; procedure) = 0
- ☐ Urine—Bilirubins; arbitrary concentration({0, 1, 2, 4}; procedure) = 2

4.8.3   **Linear differential property**: property having a magnitude expressed on a linear scale and that can be subtracted from, but cannot be divided by another property of the same kind-of-property [ref 27 modified]

Example
- ☐ Patient—Body(rectum); temperature = 37.8 °C

4.8.4   **Logarithmic differential property**: property having a magnitude expressed on a logarithmic scale and that can be subtracted from, but cannot be divided by another property of the same kind-of-property [ref 27 modified]

Example
- ☐ Patient—Urine; pH = 5.9

4.8.5   **Rational property**: property having a magnitude and that can be divided by another property of the same kind-of-property [ref 27 modified]

Examples
- ☐ Patient—Body; thermodynamic temperature = 310.5 K
- ☐ Blood—Erythrocytes; volume fraction = 0.41 = 41 %

4.9   **Quantity**: property of a phenomenon, body or substance, where the property has a magnitude that can be represented by a number and a reference [ref. 2, concept 1.1]

This subordinate concept of 'property' comprises the lower four types in Table 4.1.

NOTE 1: Metrology by definition is only concerned with 'quantity', which could simply be defined as "property having a magnitude".

NOTE 2: According to VIM3, a reference can be a measurement unit, a measurement procedure, a reference material or a combination of such [ref. 2, concept 1.1, Note 4].

4.10   **Kind-of-quantity**: aspect common to mutually comparable quantities [ref. 2, concept 1.2, but without hyphens in the term]

NOTE: In analogy to 'property' being associatively related to 'kind-of-property', the subordinate 'quantity' is associatively related to 'kind-of-quantity'.

4.11   **Generic concept systems**

It is seen that it is possible to structure at least two **generic concept systems** on the same subject field, here as tree diagrams (Figure 4.1).

The left-hand system is simple, with only one set of five coordinate-specific subordinate concepts. The right-hand system is more complicated, but incorporates the current structure in metrology as presented in VIM [ref. 2, concept 1.1], based on 'quantity'.

It is possible to insert an extra concept at the same level as 'ordinal quantity', namely 'unitary quantity', under which the last three types of quantity are

Simple two-tiered concept
system on 'property'

Four-tiered concept system
on 'property' including 'quantity'

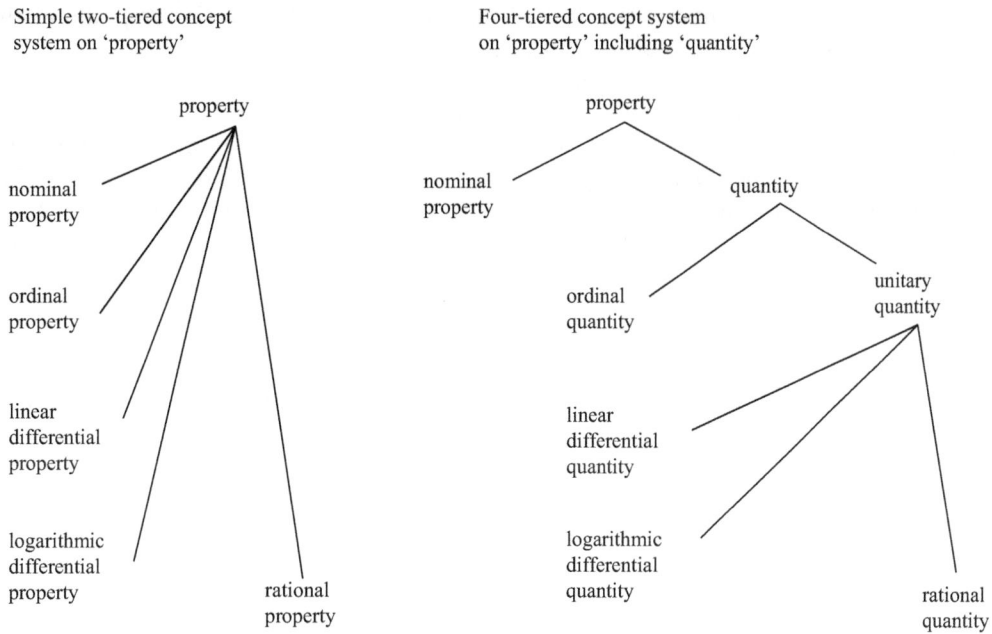

Figure 4.1 Two generic concept systems as tree diagrams.

Table 4.2 Other basic concepts involved in the production of an examination result or a measurement result.

Examination-related concepts	Measurement-related concepts
examinand	measurand
examination procedure	measurement procedure
examining system	measuring system
property value	quantity value
property value scale	quantity value scale
examination uncertainty	measurement uncertainty

coordinate-specific subordinate concepts (Figure 4.1, right-hand system). **Unitary quantity** may be defined as 'quantity with a magnitude expressed as a reference quantity multiplied by a number'.[27]

4.12 Other basic concepts

The alternative concept systems above on 'property' (Figure 4.1) can be supplemented as appropriate by other basic concepts used sequentially to arrive at a measurement result (Table 4.2).

The term "examination" and derivations have been adopted in ISO International Standards for the clinical laboratories. The term "observation" is sometimes used as a synonym of "examination", but not in this Compendium.

References

1. R. Dybkær and K. Jørgensen IUPAC–IFCC (International Union of Pure and Applied Chemistry, and International Federation of Clinical Chemistry, Commission on Clinical Chemistry). *Quantities and Units in Clinical Chemistry, including Recommendation 1966.* Munksgaard, København, 1967.
2. BIPM, IEC, IFCC, ILAC, ISO, IUPAC, IUPAP, OIML, 2012. *International vocabulary of metrology – Basic and general concepts and associated terms VIM*, 3rd edn. JCGM 200: 2012. This 3rd edition is also published as ISO Guide 99 by ISO (ISO/IEC

Guide 99-12: 2007). Replaces the 2nd edition, 1993. Available on the Web site of the BIPM: http//www.bipm.org. (acc. 2015-12).

3. M. Blombäck, R. Dybkaer, K. Jørgensen, H. Olesen and S. Thorsen, ISTH–IUPAC–IFCC (International Society on Thrombosis and Haemostasis–International Union of Pure and Applied Chemistry, Clinical Chemistry Division, Commission on Quantities and Units in Clinical Chemistry; and International Federation of Clinical Chemistry, Scientific Division, Committee on Quantities and Units). Properties and units in the clinical laboratory sciences. Part V. Properties and Units in Thrombosis and Haemostasis (Technical Report 1995), *Thromb. Haemostasis*, 1994, **71**(3), 375–394; *Pure Appl. Chem.*, 1997, **69**(5), 1043–1079; *Eur. J. Clin. Chem. Clin. Biochem.*, 1995, **33**(9), 637–660; *Clin. Chim. Acta*, 1996, **245**, S23–S28.

4. H. Olesen, D. Cowan, I. Bruunshuus, K. Klempel and G. Hill, IUPAC–IFCC (International Union of Pure and Applied Chemistry, Clinical Chemistry Division, Commission on Quantities and Units in Clinical Chemistry; and International Federation of Clinical Chemistry, Scientific Division, Committee on Quantities and Units). Properties and units in the clinical laboratory sciences. Part VI. Properties and units in IOC prohibited drugs (Technical Report 1997), *Pure Appl. Chem.*, 1997, **69**(5), 1081–1136; *Eur. J. Clin. Chem. Clin. Biochem.*, 1997, **35**(10), 805–831; *J. Chromatogr. B*, 1996, **687**, 157–182.

5. U. Forsum, H. Olesen, W. Frederiksen and B. Person, IUPAC–IFCC (International Union of Pure and Applied Chemistry, Clinical Chemistry Division, Commission on Quantities and Units in Clinical Chemistry; and International Federation of Clinical Chemistry, Scientific Division, Committee on Quantities and Units). Properties and units in the clinical laboratory sciences. Part VIII. Properties and units in clinical microbiology (Technical Report 1999), *Pure Appl. Chem.*, 2000, **72**(4), 555–745; *eJIFCC*, 2000, **12**, 1.

6. R. Cornelis, X. Fuentes-Arderiu, I. Bruunshuus and D. Templeton, IUPAC–IFCC (International Union of Pure and Applied Chemistry, Clinical Chemistry Division, Commission on Quantities and Units in Clinical Chemistry; and International Federation of Clinical Chemistry, Scientific Division, Committee on Quantities and Units)). Part IX. Properties and units in trace element (Technical Report 1997), *Pure Appl. Chem.*, 1997, **69**(12), 2593–2606; *Eur. J. Clin. Chem. Clin. Biochem.*, 1997, **35**, 833–843.

7. H. Olesen, I. Ibsen, I. Bruunshuus, D. Kenny, R. Dybkær and X. Fuentes-Arderiu, IUPAC–IFCC (International Union of Pure and Applied Chemistry, Clinical Chemistry Division, Commission on Quantities and Units in Clinical Chemistry; and International Federation of Clinical Chemistry, Scientific Division, Committee on Quantities and Units). Properties and units in the clinical laboratory sciences. Part X. Properties and units in general clinical chemistry (Technical Report 1999). *Pure Appl. Chem.*, 2000, **72**(5), 747–972.

8. H. Olesen, D. Cowan, R. de la Torre, I. Bruunshuus, M. Rohde and D. Kenny, IUPAC–IFCC (International Union of Pure and Applied Chemistry, Clinical Chemistry Division, Commission on Quantities and Units in Clinical Chemistry; and International Federation of Clinical Chemistry, Scientific Division, Committee on Quantities and Units). Properties and units in the clinical laboratory sciences. Part XII. Properties and units in clinical pharmacology and toxicology (Technical Report 1999), *Pure Appl. Chem.*, 2000, **72**(3), 479–552; *eJIFCC*, 2000, **12**, 1.

9. H. Olesen, A. Giewercman, D. M. de Krister, D. de Mortimer, H. Oshima and P. Troen, IUPAC–Int. Soc. Androl–IFCC (International Union of Pure and Applied Chemistry, Clinical Chemistry Division, Commission on Quantities and Units in Clinical Chemistry; International Society of Andrology and International Federation of Clinical Chemistry, Scientific Division, Committee on Quantities and Units). Properties and units in the clinical laboratory sciences. Part XIII. Properties and units in reproduction and fertility (Technical Report 1997), *Pure Appl. Chem.*, 1997, **69**(12), 2621–2638; *Clin. Chem. Lab. Med.*, 1998, **36**(1), 57–65; *Clin. Chim. Acta*, 1998, **271**, S5–S26.

10. I. Bruunshuus, L. K. Poulsen and H. Olesen, IUPAC–IFCC (International Union of Pure and Applied Chemistry, Clinical Chemistry Division, Commission on Quantities and Units in Clinical Chemistry; and International Federation of Clinical Chemistry,

Scientific Division, Committee on Quantities and Units). Properties and units in the clinical laboratory sciences. Endorsed by The American Academy of Allergy, Asthma and Immunology. Part XVI. Properties and units in clinical allergology (Technical Report 1999), *Pure Appl. Chem.*, 2000, **72**(6), 1067–1205; *eJIFCC*, 2000, **12**, 1.

11. P. Soares de Araujo, B. Zingales, P. Alia-Ramos, A. Blanco-Fonta, X. Fuentes-Arderiu, C. Mannhalter, S. Bojesen, I. Bruunshuus and H. Olesen, IUPAC–IFCC (International Union of Pure and Applied Chemistry, Clinical Chemistry Division, Commission on Quantities and Units in Clinical Chemistry; and International Federation of Clinical Chemistry, Scientific Division, Committee on Quantities and Units). Properties and units in the clinical laboratory sciences. Part XVIII. Properties and units for Clinical Molecular Biology (Technical Report 2004), *Pure Appl.Chem.*, 2004, **76**(9), 1799–1807.

12. K. Varming, U. Forsum, I. Bruunshuus and H. Olesen, IUPAC–IFCC (International Union of Pure and Applied Chemistry, Clinical Chemistry Division, Commission on Quantities and Units in Clinical Chemistry; and International Federation of Clinical Chemistry, Scientific Division, Committee on Quantities and Units). Properties and units in the clinical laboratory sciences. Part XIX. Properties and units for transfusion medicine and immunohematology (Technical Report 2003), *Pure Appl. Chem.*, 2003, **75**(10), 1477–1600.

13. J. Duffus, I. Bruunshuus, R. Cornelis, R. Dybkaer, M. Nordberg and W. Kuelpmann, IUPAC–IFCC (International Union of Pure and Applied Chemistry, Clinical Chemistry Division, Commission on Quantities and Units in Clinical Chemistry; and International Federation of Clinical Chemistry, Scientific Division, Committee on Quantities and Units). Properties and units in the clinical laboratory sciences. Part XX. Properties and units in environmental human toxicology (Technical Report 2007), *Pure Appl. Chem.*, 2007, **79**(1), 87–152; *eJIFCC*, 2007, **18**, 2.

14. ISO/IEC 80000-9:2009. *Quantities and units* — Part 9: *Physical chemistry and molecular physics*. Supersedes ISO 31-8: 1992.

15. W. Horwitz IUPAC-CAN 1990. Nomenclature of sampling in analytical chemistry. *Pure Appl. Chem.*, 1990, **62**(6), 1193–1208.

16. N. G. Connelly, T. Damhus, R. M. Hartshorn and A. T. Hutton, *Nomenclature of Inorganic Chemistry, IUPAC Recommendations 2005, known as the IUPAC Red Book*, RSC Publishing, Cambridge, 2005.

17. H. A. Favre and W. H. Powell, *Nomenclature of Organic Chemistry (IUPAC Recommendations and Preferred IUPAC Names 2013)*, Royal Society of Chemistry, 2013.

18. *Enzyme Nomenclature 1992* [Academic Press, San Diego, California] with Supplement 1 (1993), Supplement 2 (1994), Supplement 3 (1995), Supplement 4 (1997) and Supplement 5 (in *Eur. J. Biochem.*, 1994, **223**, 1–5; *Eur. J. Biochem.*, 1995, **232**, 1–6; *Eur. J. Biochem.*, 1996, **237**, 1–5; *Eur. J. Biochem.*, 1997, **250**, 1–6 and *Eur. J. Biochem.*, 1999, **264**, 610–650, respectively), and a www version prepared by G. P. Moss: www.chem.qmul.ac.uk/iubmb/enzyme (acc. 2015-12).

19. C. Liébecq and IUB-CEBJ, *Biochemical Nomenclature and Related Documents: A Compendium*, Portland Press, London, 2nd edn, 1992.

20. G. P. Moss IUPAC Organic Chemistry Division, Commission on Nomenclature of Organic Chemistry, Commission on Physical Organic Chemistry. *Basic terminology of stereochemistry* (Recommendations 1996). *Pure Appl. Chem.*, 1996, **68**(12), 2193–2222.

21. *International Nonproprietary Names (INN) for Pharmaceutical Substances.* — *Cumulative list 14*. WHO, Geneva, 2011.

22. *International Nonproprietary Names (INN) for Pharmaceutical Substances. Names for Radicals and Groups and Others: Comprehensive List, 2012.* WHO, Geneva, 2012.

23. ISO 1750:1981. *Pesticides and other agrochemicals — Common names*.

24. International Anatomical Committee's Nomina anatomica, histologica, cytologica (IAC 1989) Churchill Livingstone, Edinburgh, 1989.

25. International Anatomical Committee's *Nomina anatomica, histologica, cytologica and in veterinary applications* in line with the International Committee on Veterinary Gross Anatomical Nomenclature (ICVGAN 1983).

26. I. Mills, T. Cvitaš, K. Homann, N. Kallay and K. Kuchitsu, *Quantities, Units and Symbols in Physical Chemistry — The IUPAC Green Book*, RSC Publishing, Cambridge, 3rd edn, 2007.

27. R. Dybkaer, *An Ontology on property for physical, chemical, and biological systems*. 2009. http://ontology.iupac.org. (acc. 2015-12).

28. S. S. Stevens, On the theory of scales of measurement, *Science*, 1946, **103**, 677–680.

29. S. S. Stevens, Measurement, psychophysics, and utility, in *Measurement: Definition and Theories*, ed. C. W. Churchman and P. Ratoosh, John Wiley and Son, New York, 1959, p. 18–63.

SECTION 5

Principles and Practice of Kinds-of-quantity and Units

The nature of the concepts 'system' and 'kind-of-quantity' has been explained in Section 4. The aim of the present section is to develop and clarify other concepts derived from, or related to, these concepts.

5.1   Kinds-of-quantity

5.1.1   A **system of kinds-of-quantity** is a set of kinds-of-quantity, except ordinal kinds-of-quantity, together with a set of non-contradictory equations relating those kinds-of-quantity. This definition is analogous to that of 'system of quantities' in the VIM [ref. 1, concept 1.3]. A kind-of-quantity of a given system of kinds-of-quantity may be a **base kind-of-quantity** or a **derived kind-of-quantity**. The system of kinds-of-quantity based on the seven base kinds-of-quantity: length, mass, time, electric current, thermodynamic temperature, amount of substance and luminous intensity is termed the **International System of Quantities** or **ISQ**. The definition of this system is analogous to that of VIM [ref. 1, concept 1.6].

5.1.2   In parallel with the above concepts, some concepts are defined around the concept of **measurement unit**. Thus, a **system of units** is a set of **base units** and **derived units**, together with their multiples and submultiples, defined in accordance with given rules, for a given system of kinds-of-quantity.[1] The **International System of Units** or **SI** is a system of units based on the International System of Quantities, their terms and symbols, including a series of prefixes and their terms and symbols, together with rules for their use, adopted and maintained by the General Conference on Weights and Measures (CGPM). These definitions are analogous to those of VIM [ref. 1, concepts 1.13, 1.16].

5.1.3   The first records of human history from about 10 000 years ago give numbers of livestock. Around 5000 BC, the first writings on a flat surface from the Mesopotamian basin are book-keeping records of land area, crop yields and amounts of foodstuffs. The records from Sumeria, Babylon and Palestine indicate the concern for calibration, quality assurance and/or agreement on units and measures. However, until the 18th century, the diversity of systems of units and measures, even between regions of one country, were a hindrance to trade. The French Revolution provided the impetus for a scientifically based system of units, the metric system. During the 19th century, the metre and kilogram became the usual units, except in Anglo-Saxon countries.

Compendium of Terminology and Nomenclature of Properties in Clinical Laboratory Sciences: Recommendations 2016
Edited by Georges Férard, René Dybkaer and Xavier Fuentes-Arderiu
© International Union of Pure and Applied Chemistry 2017
Published by the Royal Society of Chemistry, www.rsc.org

In 1875, 17 countries united in the **Metre Convention** with the purpose of agreeing on definitions of the units and of distributing related standards throughout the world. The legislative body of the Convention is the **General Conference on Weights and Measures** with representatives of 57 countries in 2015. It meets every 4 years. The **International Committee for Weights and Measures** (CIPM) meets every year, supervising specialist committees and preparing draft resolutions for the General Conference.

With the rapid evolution of science and technology in the 19th century, diversity of units increased again, but between disciplines rather than between regions of the world. For instance, energy rate (or power) was expressed in watts, kilocalories per hour, horsepower, ergs per second, British thermal units per hour, or calories per minute. The diversity of units of pressure, force and work became a source of confusion. Some quantities were expressed on technical scales related to the method of measurement, for instance degrees Soxhlet, degrees Baumé, degrees Engler and Redwood seconds. Time was wasted on conversions and comparisons. Misunderstandings were rife.

One of the few areas where there has been little unification in reporting of results is on units of time. Proposals to switch to a decimal system of measuring time stranded on the awkward relationship of day, month and year. Essentially we are still using the system developed by the Babylonians over 3000 years ago.

About 1900, two coherent systems of units were developed. Giorgi developed a system based on the metre, kilogram, second and an electrical unit (**Giorgi system**); others preferred a system based on the centimetre, gram and second (**CGS system**). Giorgi first chose the ohm as the electrical unit, but later the IEC chose the ampere as the electrical unit. This system was known as the MKSA system. In 1960, an extension of the system developed by Giorgi was recognized as the **International System of Units (SI)**, extended in 1971 to include the mole.

5.1.4   A major reason for choosing a coherent system of units is the complexity of calculations with conversion factors, especially between units related by factors other than powers of 10 (*e.g.* second, minute, hour, day), and the extra risk of misunderstandings.

5.1.5   To avoid the haphazard factors in relationships between numerical values, equations should be based on quantities (see §5.6.5) and not on numerical values unless every quantity in an equation is first expressed in its coherent SI unit.

5.1.6   Most kinds-of-quantity can be defined from one another, by multiplication or division with or without factors (§5.6.2). A few may be chosen arbitrarily as **base kinds-of-quantity**. A base kind-of-quantity is one that is conventionally accepted as independent of other base kinds-of-quantity in a system of kinds-of-quantity. The rest are **derived kinds-of-quantity**. A derived kind-of-quantity is one that is defined as a function of the base kinds-of-quantity of a given system of kinds-of-quantity.

5.1.7   By choosing a **base unit** for each base kind-of-quantity, coherent units for all other kinds-of-quantity, derived units, can be defined in terms of the base units in such a way that factors between units are avoided. Equations between numerical values then have the same form as the equations between kinds-of-quantity. A system of units defined in this way is said to be **coherent** with respect to the given kinds-of-quantity and equations between them. Table 5.1 lists the choices of base kinds-of-quantity and base units for the **International System of Quantities (ISQ)** and the **International System of Units (SI)**.

5.1.8   The choice of base unit for each base kind-of-quantity is a matter of practical convenience.

Table 5.1   Base kinds-of-quantity, base units and their dimensional symbols (§5.2) in the International System of Units (SI). The status of number of entities is also that of a base kind-of-quantity [ref. 1, concept 1.16, Note 4].

Base kind-of-quantity		Base unit		Dimension
Term	Symbol	Term	Symbol	Symbol
number (of entities)	N	one	1	1
length	l	metre	m	L
mass	m	kilogram	kg	M
time	t	second	s	T
electrical current	I, i	ampere	A	I
thermodynamic temperature	T, θ	kelvin	K	Θ
amount-of-substance	n	mole	mol	N
luminous intensity	I_v	candela	cd	J

5.1.9   Currently the CIPM, advised by the Consultative Committee for Units (CCU), is considering a general revision of the definitions of the seven base units of the SI so that each unit will be defined indirectly by specifying explicitly an exact value for a well-recognized fundamental constant. The format of an explicit-constant definition could be "the [unit term, symbol], is equal to exactly a fixed numerical value of the unit of [kind-of-quantity]; its magnitude is set by fixing the numerical value of the [fundamental constant] to be equal to exactly [numerical value] when it is expressed in the unit [SI coherent derived unit]". Some of the advantages of such an approach are that the last material artefact defining a base unit, the slightly unstable prototype of the kilogram, becomes an historic item only and that each base unit would have an exact value without measurement uncertainty. A disadvantage for some of the definitions might be that the implicit physical relation between a fundamental constant and the corresponding base unit is not easily discerned.

5.2   Dimension of a Kind-of-quantity

For dimensional analysis, each base kind-of-quantity of ISQ is assigned a **dimension** represented by a sans-serif capital letter symbol corresponding to the letter symbol for the kind-of-quantity, except for number (of entities) and for luminous intensity (Table 5.1).

Example
☐ The dimension of mass is represented by M (Table 5.1).
The dimension of any derived quantity can then be expressed as a product of powers of those seven dimensions: $\mathsf{L}^{\alpha}\mathsf{M}^{\beta}\mathsf{T}^{\gamma}\mathsf{I}^{\delta}\Theta^{\varepsilon}\mathsf{J}^{\xi}\mathsf{N}^{\eta}$.

Examples
☐ The derived kind-of-quantity 'mass concentration' has the dimension $\mathsf{L}^{-3}\mathsf{M}^{1}\mathsf{T}^{0}\mathsf{I}^{0}\Theta^{0}\mathsf{J}^{0}\mathsf{N}^{0} = \mathsf{L}^{-3}\,\mathsf{M}$ and the coherent SI unit kg m^{-3}.
☐ The derived kind-of-quantity 'force' has the dimension $\mathsf{L}^{1}\mathsf{M}^{1}\mathsf{T}^{-2}\mathsf{I}^{0}\Theta^{0}\mathsf{J}^{0}\mathsf{N}^{0} = \mathsf{LMT}^{-2}$ and the coherent unit kg m s^{-2}.

5.3   Kinds-of-quantity of Dimension One

5.3.1   If the exponents of the dimensional product are all zero, the kind-of-quantity is said to be of **dimension one** or (less correct) **dimensionless**: $\mathsf{L}^{0}\mathsf{M}^{0}\mathsf{T}^{0}\mathsf{I}^{0}\Theta^{0}\mathsf{J}^{0}\mathsf{N}^{0} = 1$.

Examples
☐ number of entities of component (Table 5.1; §8.1.1), dim $N_B = 1$
☐ substance fraction of component (§8.7.1), dim $x_B = 1$
☐ relative kinds-of-quantity such as relative volumic mass of system (§8.4.3), dim $d_1 = 1$

For such kinds-of-quantity, the coherent unit in any system of units is the unit 1, often termed "unity". According to ISO/IEC 80000-1,[2] the unit one, symbol 1, is generally not written explicitly in the value of such a quantity.

5.3.2 CGPM[10] only recognizes units of plane angle and solid angle. ISO/IEC 80000-1[2] and CGPM accept the symbols per cent (%) and per mille (‰) to represent the numbers 0.01 and 0.001, respectively. The use of part(s) per thousand (ppt), part(s) per million (ppm) for 10^{-6}, part(s) per hundred million (pphm) for 10^{-8}, parts per billion (ppb) for either 10^{-9} or 10^{-12} is discouraged. They should be replaced by unambiguous units: ppm should be replaced by μmol mol^{-6} or mg kg^{-6} or by 10^{-6}, as the case may be.

5.3.3 The expression of very large and very small values of quantities of dimension one remains a problem and source of misunderstanding (*e.g.* in tables and graphs, §5.11.3). The decimal prefixes of Table 5.3 cannot be used to form multiples or submultiples of the unit one, until a special symbol is agreed internationally. Discussions in IUPAC, ISO, IEC and CGPM have not reached a consensus on a possible symbol nor on alternative modes of expression of very large or very small values of quantities of dimension one. In Section 8, powers of 10^{3n} or redundant units, *e.g.* mL L^{-1} (§8.3U) and g kg^{-1} (§8.4U), are offered as alternatives. ISO/IEC 80000-1[2] gives powers of 10.

Examples

☐ A mass fraction of 0.000 000 005 × 1 may be expressed as 5×10^{-9}, $5 \cdot 10^{-9}$ or 5 μg kg^{-1}, but not as '5 n1' ('five nano-ones').

☐ Blood—Erythrocytes; number concentration $= 5 \times 10^{12}$ L^{-1}, but not 5 T1 L^{-1} ('five tera-ones per litre')

Table 5.2 Derived units of the International System of Units (SI) with special terms or symbols. The sequence of the list is as in Sections 8 and 9, essentially by order of dimension (Table 5.1) and increasing powers of those dimensions, first positive and then negative. For electrical and luminous kinds-of-quantity, systematic nomenclature of the kinds-of-quantity is based on electrical charge (unit C = A s) and amount-of-light (unit lm s = cd sr s). The term and symbol for the katal have been recognized by IUPAC and IFCC,[6] IUB,[7] WHO,[8] and finally by CGPM (1999, Resolution 12). Note that the symbols t, ϑ and Φ have several meanings in Table 5.2, those of Φ being distinguished here by subscripts. N (Table 5.1) is also to be distinguished from N (this table).[a]

Kind-of-quantity		Unit		Definition in
Term	Symbol	Term	Symbol	SI base units
plane angle	$\alpha, \beta, \gamma, \vartheta$	radian	rad	m m^{-1} = 1
solid angle	Ω	steradian	sr	m^2 m^{-2} = 1
frequency	f, ν	hertz	Hz	s^{-1}
activity referred to a radionuclide	A	becquerel	Bq	s^{-1}
massic energy of ionizing radiation, absorbed dose	D	gray	Gy	m^2 s^{-2}
massic energy equivalent of ionizing radiation absorbed, dose equivalent	H	sievert	Sv	m^2 s^{-2}
force	F	newton	N	kg m s^{-2}
pressure, stress	p	pascal	Pa	kg m^{-1} s^{-2}
energy, work, amount-of-heat	E, Q_e	joule	J	kg m^2 s^{-2}
power, energy rate	P, Φ_e	watt	W	kg m^2 s^{-3}
electrical charge	Q	coulomb	C	A s
magnetic flux	Φ_m	weber	Wb	kg m^2 A^{-1} s^{-2}
magnetic flux density	B	tesla	T	kg A^{-1} s^{-2}
electric potential difference, electric tension	U, V	volt	V	kg m^2 A^{-1} s^{-3}
inductance	L	henry	H	kg m^2 A^{-2} s^{-2}
electrical conductance	G	siemens	S	A^2 s^3 kg^{-1} m^{-2}
electric resistance	R	ohm	Ω	kg m^2 A^{-2} s^{-3}
capacitance	C	farad	F	A^2 s^4 kg^{-1} m^{-2}
Celsius temperature	t, ϑ	degree Celsius	°C	K
light rate, luminous flux	Φ_v	lumen	lm	cd sr
illuminance, areic light rate	M_v	lux	lx	cd sr m^{-2}
catalytic activity	z	katal	kat	mol s^{-1}

[a]⟨19⟩ Activity referred to a radionuclide is sometimes incorrectly termed radioactivity.

5.4 Nomenclature of Kinds-of-quantity

5.4.1   This section summarizes several attempts within IUPAC and ISO to create a logical descriptive system of nomenclature for kinds-of-quantity and some of the problems with existing terms.

5.4.2   As pointed out in §5.1.6, kinds-of-quantity belong to one of the two fundamental groups: base kinds-of-quantity, Q, and derived kinds-of-quantity, q. Under a different point of view, kinds-of-quantity may be extensive or intensive. An **extensive kind-of-quantity** is that used in an individual quantity whose value is dependent on the size of the system, whereas an **intensive kind-of-quantity** is that used in an individual quantity whose value is independent of the size of the system. Base kinds-of-quantity are extensive kinds-of-quantity, except for time and thermodynamic temperature, but some derived kinds-of-quantity are also extensive kinds-of-quantity (*e.g.* area, volume). Regarding derived kinds-of-quantity, many of them are intensive.

5.4.3   For derived kinds-of-quantity involving division, the defining kinds-of-quantity in the numerator and the denominator of the definition may be of the same kind-of-quantity (§5.4.4) or of different kinds-of-quantity (§5.4.5 onwards).

5.4.4   If two **defining kinds-of-quantity are identical, and one is divided by the other,** the term of the derived kind-of-quantity, q, includes the term for the defining kind-of-quantity, Q, omitting the words **amount-of** (as appropriate), and one of the following modifiers (§5.4.4.1–5.4.6.14). Some obsolete or discouraged synonyms for these kinds-of-quantity can be found in Section 8.

5.4.4.1   **fraction:** quotient of two identical kinds-of-quantity, for which the numerator kind-of-quantity relates to a component B and the denominator kind-of-quantity relates to the system 1, Q_B/Q_1 or $Q_B \bigg/ \left(\sum_A^N Q_i \right)$ or $Q_B \bigg/ \left(\sum_B^N Q_i \right)$, where B $\subseteq$ 1 and summation is usually for all components A to N in a gaseous system and B to N in a solution

  Examples
  - □ number fraction of people with anaemia: number of people with anaemia divided by total number of people (in a specified population) (see also §8.2.1.4)
  - □ Haemoglobin beta chain (B)-*N*-(1-deoxyfructos-1-yl)haemoglobin beta chain; subst.fr. $= 51$ mmol mol$^{-1} = 0.051 = 5.1$ %

5.4.4.2   **ratio:** quotient of two identical kinds-of-quantity, for which the numerator kind-of-quantity relates to a component B and the denominator kind-of-quantity to another component of the same system, commonly treated as a reference component, $(Q_B/Q_R)_1$

  Examples
  - □ volume ratio (§8.3.2.1): a volume divided by another volume, V_B/V_R
  - □ mass–concentration ratio: a mass divided by a volume, divided by another mass and a volume. If the volumes are the same, it can be more simply termed a mass ratio (§8.4.2), $(\rho_B/\rho_R)_1 = (m_B/m_R)_1$

5.4.4.3   **relative:** quotient of two identical kinds-of-quantity, commonly two kinds-of-quantity related to different systems, the second being a reference system Q_1/Q_0

  Example
  - □ relative time (§8.5.1): a time divided by another time

5.4.5   For derived kinds-of-quantity defined from equations of the type $q = Q/x$, where x may represent an extensive or intensive kind-of-quantity, the term for the derived kind-of-quantity, q, should include the term of the numerator kind-of-quantity, Q, and a distinct word, usually an adjective, designating the **denominator kind-of-quantity**, x. In English, the adjectival ending -ic provides distinction from other adjectival forms, which often have other meanings. Such term formation is also mentioned in ISO/IEC 80000-1[2] and in the IUPAC 'Green Book',[3] see §5.4.6.

    Examples
- ☐ lineic
- ☐ volumic
- ☐ massic
- ☐ areic

5.4.6   The following terms designate **denominator kinds-of-quantity**, x. Some obsolete or discouraged synonyms for these kinds-of-quantity can be found in Section 9.

5.4.6.1   **entitic**: numerator kind-of-quantity, usually extensive, divided by number of entities, Q_B/N_B
    Example
- ☐ entitic mass, m_B/N_B (§9.8.2)

5.4.6.2   **steradic**: numerator kind-of-quantity of a component divided by solid angle through which the component is projected
    Example
- ☐ steradic energy rate, Φ_e/Ω (§9.48.2)

5.4.6.3   **lineic**: numerator kind-of-quantity, usually extensive, divided by length, Q/l
    Example
- ☐ lineic electrical conductance, G/l (§9.63.1)

5.4.6.4   **gradient**: differential change in a local intensive kind-of-quantity with position divided by change in position, **grad** q (or ∇q) $= \mathrm{d}q_{x,y,z}/\mathrm{d}(x,y,z)$
    Example
- ☐ number concentration gradient, **grad** $C_B = \nabla C_B = \mathrm{d}C_B/\mathrm{d}(x,y,z)$ (§9.7.1)

5.4.6.5   **areic**: numerator kind-of-quantity, usually extensive, divided by area, Q/A
    Example
- ☐ areic volume rate, Φ_v/A (§9.20.2)

5.4.6.6   **volumic**: numerator kind-of-quantity of system divided by volume of system, Q_1/V_1
    Example
- ☐ volumic mass, m_1/V_1 (§9.14.1)

5.4.6.7   **concentration**: numerator kind-of-quantity related to some amount of component divided by volume of system, Q_B/V_1
    NOTE: Concentration must be specified (amount-of-substance, mass, number, ...) when necessary.
    Examples
- ☐ mass concentration of a component B in a system 1 (§9.14.2): mass of component divided by volume of the system
- ☐ substance concentration of a component B in a system 1 (§9.88.1) (rather than amount-of-substance concentration §5.4.14): amount-of-substance of component divided by volume of the system

5.4.6.8 **massic**: numerator kind-of-quantity of system divided by mass of system, Q_1/m_1
   Example
   □ massic volume, V_1/m_1 (§9.15.1)

5.4.6.9 **content**: numerator kind-of-quantity related to some amount of component divided by mass of system, Q_B/m_1
   Example
   □ substance content, n_B/m_1 (§9.91.1)

5.4.6.10 **rate**: change in numerator kind-of-quantity divided by change in time, dQ/dt
   Example
   □ areic substance rate, $(dn_B/dt)/A$ (§9.95.1)

5.4.6.11 **kelvic**: change in numerator kind-of-quantity divided by change in thermodynamic temperature, dQ/dT (*e.g.* §9.72.1)

5.4.6.12 **molar**: numerator kind-of-quantity due to a component or system divided by amount-of-substance of that component or system Q_B/n_B or Q_1/n_1
   Example
   □ molar mass, (m_B/n_B) (§9.92.1)
NOTE: This commonly accepted meaning of "molar" is also given in ISO/IEC 80000-1, Annex A.6.5, but with a note that "the term "molar" violates the principle that the name of the quantity shall not reflect the name of a corresponding unit (in this case, mole)".[2] This Compendium uses "molar" in the sense given above.

5.4.6.13 **partial molar**: change in a numerator kind-of-quantity due to addition of a component divided by amount-of-substance of that component, $\partial Q_B/\partial n_B$
   Example
   □ partial molar volume of a component B, $V_{m,B}$ (§9.89.1) for change in volume due to a component divided by change in amount-of-substance of that component, $\partial V_B/\partial n_B$

5.4.6.14 In this Compendium, furthermore, a distinction is made between the adjectival form **electrical**, used when the definition of the kind-of-quantity has electrical charge (unit coulomb) in the numerator, and **electric** indicating that electrical charge is in the denominator.
   Example
   □ electrical conductance, unit siemens$=S=s$ C^2 $kg^{-1}m^{-2}$ and electric resistance, unit ohm$=\Omega=$ kg m^2 C^{-2} s^{-1}

5.4.7 The systematic terms are given under tag ⟨2⟩ in the lists of kinds-of-quantity in Sections 8 and 9 alongside current unambiguous (and recommended) terms and synonyms (in bold type, under tag ⟨3⟩).

5.4.8 Besides systematic terms, many **traditional terms** exist. Some of these terms indicate a **process**, observed or assumed (§4.4), associated with the kind-of-quantity. If there is a local or disciplinary convention, the meaning is then implicit. Descriptive terms are, however, preferable, particularly if the disciplinary convention is not widely known. In the following examples, terms printed in bold characters are recommended, and preferred to those in italic type as explained in §3.2. Terms printed in ordinary type are accepted when specifications are not necessary.
   Examples
   □ **elongation**, in which something elongates, for change in length, Δl (§9.1.1.1)

☐ **acceleration,** for rate of change of linear velocity, $a = \mathrm{d}v/\mathrm{d}t = \mathrm{d}^2x/\mathrm{d}t^2$ (§9.34.1)

☐ absorption, for an interaction of electromagnetic radiation with matter and for the energy fraction of radiation absorbed into a system

☐ absorption also for absorption of matter and for an amount-of-substance or mass of a component absorbed

☐ *flow* for movement of a liquid and for the amount (commonly volume or mass) of liquid flowing, Q, or for the **volume rate,** or **mass rate,** $\mathrm{d}Q/\mathrm{d}t$ (§9.26.1)

☐ *rate of growth* for **number rate, length rate, volume rate** or **mass rate** (§9.26.1)

5.4.9   Sometimes the word ending **-ion** designating a process, has been replaced by **-ance** or **-ivity,** with a tendency for **-ivity** to refer to a more complex coefficient. Sometimes **-ability** (or **-ibility**) was used with a general meaning of ability or 'power' or tendency to undergo a process. Often terms ending in **-ance** refer to kinds-of-quantity of dimension one.

Examples

☐ absorbance for a logarithmic kind-of-quantity describing absorption of radiation (§8.9.1)

☐ absorptance for energy fraction of a radiation absorbed (§8.6.1.1)

☐ absorptivity (or absorption coefficient) for absorbance divided by path length traversed by the radiation (§9.2.2.4) or by path length and mass concentration (§9.13.4) or substance concentration (§9.88.1). The term is also used for absorption of solutes.

☐ electrophoretic mobility for the tendency of a component to move in an electrical field (§9.55.1)

5.4.10   The words **factor** and **coefficient** in the term for a kind-of-quantity tell something of how the kind-of-quantity is used in equations of proportionality. **Factor** is reserved for multipliers of dimension 1. **Coefficient** is reserved for other multipliers.[2] More specific terms can be constructed by means of the words in §5.4.4 and §5.4.6.

NOTE: Absorption factor and transmission factor, respectively, were also termed *coefficients* in early editions of the recommendations of IUPAP and ISO/IEC 80000-1.[2] They can be termed **energy fractions** of a radiation absorbed and transmitted.

Constant should be reserved for universal constants of nature.

Examples

☐ Avogadro constant (§9.83.1)

☐ equilibrium constant (§8.9.6, §9.41.8 and §9.88.4)

5.4.11   **Qualifiers** are commonly added to terms of the types described in §5.4.11–12 to distinguish between different kinds-of-quantity.

Examples

☐ *linear* absorptivity; linear absorption coefficient → lineic absorbance: absorbance divided by path length

☐ *specific* absorptivity; specific absorption coefficient → massic area absorbance: absorbance divided by path length and by mass concentration of the absorbing component

☐ *molar* absorptivity; molar absorption coefficient → molar area absorbance: absorbance divided by path length and substance concentration of the absorbing component

5.4.12   Qualifiers in traditional terms may indicate a defining kind-of-quantity (*e.g.* linear for length) but fail to distinguish whether the defining kind-of-quantity is part of the numerator or denominator of the definition. Sometimes such terms do not mention all defining kinds-of-quantity.

Examples
- ☐ linear density→lineic mass (denominator length), m/l (§9.10.1)
- ☐ mass density→volumic mass (numerator mass), $\rho_1 = m_1/V_1$ (§9.14.1)
- ☐ mass absorption coefficient (denominator mass but no indication of length or volume)→massic area absorbance $A_{abs}/(l\ \rho) = A_{abs}/[l(m/V)] = A_{abs}A_{area}/m$

5.4.13   In some current terms, the word representing the numerator kind-of-quantity is converted to adjectival form, but that practice should be limited because the adjectival form, like many other adjectives, may refer to either a numerator or denominator kind-of-quantity (§5.4.12).

Examples
- ☐ thermal (or heat) capacity→kelvic enthalpy: enthalpy (or heat) divided by a temperature (§9.76.1)
- ☐ electrical current→electricity rate: electrical charge divided by time (§9.52.1)
- ☐ molar energy: energy divided by an amount-of-substance (§9.103.1)
- ☐ molar absorption coefficient→molar area absorbance (§9.86.3): absorbance divided by absorption path length and by substance concentration (*i.e.* multiplied by area and divided by amount-of-substance $A_{abs}/(l\ c_B) = A_{abs}A_{area}/n_B$ (Similar in usage and meaning to the trivial term are molar optical rotatory power, §9.86.2, and molar electrical conductivity, §9.106.1).

5.4.14   Of special concern to clinical laboratory professionals was the term of **substance concentration** (§9.88.1). In the 1960s, the terms normality, equivalent and molarity were discouraged by draft recommendations of ISO and IUPAC. After discussions with IUPAC-IDCNS at Madrid in 1975 (Section 1) the shorter term *substance concentration* was agreed for clinical laboratory sciences. The Green Book uses the term *amount concentration*, but mentions the preference of this Compendium.[3]

5.5   Symbols for Kinds-of-quantity

5.5.1   Symbols for kinds-of-quantity are used in many disciplines.[4] The IUPAC has published a recommendation on these symbols,[3] which are not primarily intended for use in running text but rather for use in **equations, tables** and **figures**; if needed in text, their form is invariable. Rules and usages can be summarized as follow: Symbols for kinds-of-quantity are usually **single letters**, capital or lower case, of the Latin or Greek alphabet printed in italic (sloping) script (irrespective of the letter type used in surrounding text). Greek letters should be of a form distinct from those used for unit symbols (§5.7). **Vector kinds-of-quantity** may be indicated by italic bold face letters (preferred here) or by italic and a superior right-facing arrow (not used here). A symbol for a kind-of-quantity may be modified by a **subscript** or **parenthetic symbol**. Either type of **modifier** is italic if it also represents a kind-of-quantity; otherwise it is printed in roman (upright) letters. A modifier may serve to distinguish different kinds-of-quantity with the same symbol or to provide information about the component or system. Recommended symbols for different kinds-of-quantity are listed in Sections 8–9. There are far more kinds-of-quantity than letters of the Roman and Greek alphabets, and so meanings must always be defined.

5.5.2   However, IUPAC and IFCC have recommended **abbreviations** of kinds-of-quantity rather than symbols for describing quantities in clinical laboratory sciences in a short format. A paper with abbreviations in many languages for the most usual kinds-of-property was published by IUPAC and IFCC[5] (see Table 6.2 of this Compendium).

5.5.3   Exceptionally, the designation "pH" is both a term and a symbol for a kind-of-quantity and is always printed in upright type.

5.6   Mathematical Operations with Quantities

5.6.1   Unitary quantities cannot be meaningfully **added** to or **subtracted** from one another unless they are of the same kind, *i.e.* they must employ the same kind-of-quantity. Furthermore, their numerical values can only be added or subtracted if the units are identical:

$$x_1 + x_2 = (\{x_1\} \times [x_1]) + (\{x_2\} \times [x_2])$$

Only if the unit $[x_1] = [x_2] = [x]$
then

$$x_1 + x_2 = (\{x_1\} + \{x_2\}) \times [x]$$

5.6.2   Quantities can be **multiplied** or **divided** to give new quantities according to the rules of algebra. The resulting equations are independent of the units chosen. For the ways of representing these operations, see §5.12.

   Example
   □ volume of solution 1, V_1 $= 20\ \text{mL} = 0.020\ \text{L}$
   mass of component B, m_B $= 3\ \text{g}$
   mass concentration of a component B, ρ_B $= m_B/V_1$
                                                                            $= (3\ \text{g})/(0.020\ \text{L})$
                                                                            $= [3/(0.020)]\ \text{g L}^{-1}$
                                                                            $= [(3000)/20]\ \text{g L}^{-1}$
                                                                            $= 150\ \text{g L}^{-1}$

5.6.3   Not all **arithmetic** or **algebraic operations** that are possible with numbers are possible with all types of quantity. Because they can be multiplied and divided, unitary quantities can also be **raised to a power**. However, one cannot derive an **exponential**, a **logarithmic**, or a **trigonometric function** of a quantity. That is only possible for **numbers, numerical values**, and **quantities of dimension one**. Ordinal quantities cannot be multiplied or divided.

5.6.4   A function of a certain quantity, for instance with respect to a spectral quantity (§7.3.13–14) or to time may be defined either as a **differential (or derivative) function** at a position or time or as a **cumulative (or rising) quantity** over an interval. The differential function is denoted by a subscript and the cumulative function by a parenthesis.

   Example
   □ The wavelength differential function of energy at a given wavelength, $dQ_e(\lambda)/d\lambda$, is symbolized $Q_{e,\lambda}$ and the wavelength cumulative function of energy over a given wavelength interval $\Delta\lambda$, $\int Q_{e,\lambda}d\lambda$, as $Q_e(\Delta\lambda)$.
   A **differential function** is expressed in a unit of the quotient, whereas the **cumulative function** is expressed in a unit of the initial quantity.

   Example
   □ $Q_{e,\lambda}$ has the SI unit joule per metre; $Q_e(\Delta\lambda)$ has the SI unit joule.

5.6.5   Two types of equation have been used in science. The first is **equations between numerical values**, which are discouraged (ISO/IEC 80000-1[2]). The second is **equations between quantities** in which **mathematical relations between quantities in a given system of quantities** are independent of **measurement units**[1].

The numerical factors in equations between numerical values vary with the choice of units, whereas those in equations between quantities are constant. Algorithms with numerical values are widely used in computer programming and result in "opacity" of the algorithms if errors need to be traced in factors or coefficients used in equations.

Example

☐ **Relationship between numerical values** (discouraged)

If n is amount-of-substance (mmol) and V is volume (L),

$c = n/V$

so that if $n = 1$ and $V = 0.2$

$c = 1/0.2 = 5$

☐ **Relationship between quantities** (recommended)

If n is amount-of-substance and V is volume,

$c = n/V$

so that if $n = 1$ mmol and $V = 0.2$ L,

$c = (1 \text{ mmol})/(0.2 \text{ L})^{-1} = 5 \text{ mmol L}^{-1}$
$= 5 \times 10^{-6} \text{ mol mL}^{-1} = 5 \times 10^{-3} \text{ mol L}^{-1}$

As long as the symbol c is defined as a quantity (right column), the calculation and the result of calculation can be expressed in any units of amount-of-substance, volume and substance concentration, even though the numerical values change.

5.7 Symbols and Terms for Units

5.7.1 **Measurement units** are represented by symbols of one, two or three letters in an upright lower-case letter except if the unit is named after a person. Its term is still written with an initial lower-case letter; only its symbol has an initial capital. An exception is the alternative symbol L for **litre** (Table 5.4), which avoids confusion with the numeral one (1) and with the modulus sign (|). If the lower-case letter l be preferred, a typeface must be used that clearly distinguishes the letter l from the numeral 1. As a unit symbol, L has no other possible connotation except formerly the lambert (L or La), a unit of areic light rate (§9.80.1). Rules for writing of unit prefixes are given in §5.9.4 and Table 5.2.

Table 5.3 SI prefixes denoting decimal factors, 10^n. Da, Danish; Es, Spanish; Gr, Greek; It, Italian; La, Latin; No Norwegian. m, exponent of 10^3.

Term	Symbol	n	m	Mnemonic
yotta	Y	24	8	La octo (8)
zetta	Z	21	7	La septem (7)
exa	E	18	6	Gr hexa (6)
peta	P	15	5	Gr penta (5)
tera	T	12	4	Gr (monster)
giga	G	9	3	Gr gigas (giant)
mega	M	6	2	Gr megas (great)
kilo	k	3	1	Gr chilioi (1000)
hecto	h	2		Gr hekaton (100)
deca	da	1		Gr deka (10)
deci	d	−1		La decem (10)
centi	c	−2		La centum (100)
milli	m	−3	−1	La mille (1000)
micro	μ[a]	−6	−2	Gr mikros (small)
nano	n	−9	−3	La nanus (dwarf)
pico	p	−12	−4	Es pico (small amount) It piccolo (small)
femto	f	−15	−5	Da, No femten (15)
atto	a	−18	−6	Da, No atten (18)
zepto	z	−21	−7	La septem (7)
yocto	y	−24	−8	La octo (8)

[a]In fonts without Greek letters, u is acceptable if an explanation is given.

Table 5.4 Non-SI units accepted for use together with the International System of Units (SI).[10] For experimentally obtained values of dalton and electronvolt, standard measurement uncertainty (*u*) is stated. The unified atomic mass unit (u) is recognized by CGPM but the term dalton and symbol Da are preferred by IUPAC-IUB;[11] results can however alternatively be expressed as molar mass (§9.8U, ⟨19⟩ (3)). Other units are recognized by CGPM[10] and mentioned in the SI brochure [ref. 10, Table 6]; those marked here with an asterisk (*) are mentioned in the SI brochure [ref. 10, Table 6] as in 'common everyday use' and are in national legislation of most countries.

Unit		
Term	Symbol	Value in coherent SI unit
degree (of arc)	°	$= (\pi/180)$ rad
minute (of arc)	′	$= (\pi/10\,800)$ rad
second (of arc)	″	$= (\pi/648\,000)$ rad
litre	L, l	$= 0.001$ m$^3 = $ dm^3
tonne	t	$= 1000$ kg
dalton	Da	$\approx 1.660\,538\,921$ yg; $u_c = 0.000\,001\,073$ yg^a
*day	d	$= 86\,400$ s
*hour	h	$= 3600$ s
*minute	min	$= 60$ s
electronvolt	eV	$\approx 160.217\,656\,5$ zJ; $u_c = 0.000\,003\,5$ zJ

^{a}Recommended value.[12]

5.7.2 Greek letters give some technical problems. For unit symbols (and elementary entities) they should, if possible, be printed in an upright letter, whereas a symbol for a kind-of-quantity should be printed as a sloping letter. For certain Greek letters, there are stylized and cursive variants (*e.g.* lower-case phi ϕ and φ). The cursive variant is preferred as designation of a kind-of-quantity.

Examples
- □ µg for microgram (Table 5.2) but μ for linear napierian attenuation coefficient (§9.2.2.1)
- □ the symbol Ω for ohm (Table 5.4) but Ω for solid angle
- □ Φ for photon but Φ_e for energy rate (§9.48.1)

5.7.3 The form of **unit symbols** does not change with textual grammar. They do not add a plural ending, nor change between lower case and capital. In contrast to abbreviations, they are written without a final full stop (except at the end of a sentence, if used in text), because a full stop may have distinct meaning in symbolic language. By contrast, the form and usage of plurals of **terms for units** follow the rules of each language.

Example
- □ Five grams is symbolized 5 g, not 5 g., or 5 gs, or 5 G.

5.8 Coherent Derived Units of the International System of Units

5.8.1 Most coherent units of derived kinds-of-quantity are expressed in symbols representing the multiplication or division or both of the component base units and are termed **compound units**.

Example
- □ Mole per square metre second (mol m^{-2} s^{-1}) is the SI-coherent compound unit of the kind-of-quantity areic substance rate (§9.95.1).

5.8.2 Certain derived units of the International System have **special terms** or **symbols**, usually both (Table 5.4).

5.9 Multiples and Submultiples of Units

5.9.1   If a system of units is chosen with only one unit for each dimension, there are bound to be **very large** and **very small numerical values** for some individual quantities.

> *Examples*
> ☐ Avogadro constant, N_A or $L \approx 602\,214\,129\,000\,000\,000\,000\,000\,000$ mol^{-1}
> ☐ rest mass of an electron,
> $m_e \approx 0.000\,000\,000\,000\,000\,000\,000\,000\,000\,000\,910\,938$ kg

5.9.2   In historical systems of units, such cumbersomely large or small numerical values were avoided by choosing a set of units differing by **arbitrary** but usually **integral factors**.

> *Example*
> ☐ In the avoirdupois system masses are expressed, in order of increasing mass, in the units grain, drachm, ounce, pound, stone, hundredweight and ton, related sequentially by factors of 437.5 (grain to ounce), 16 (drachm to ounce), 16, 14, 8 and 20.

5.9.3   In the historical metric system, factors were limited as far as possible to **decimal factors**, *i.e.* factors of ten to an integral power n, 10^n.

> *Examples*
> ☐ land area in the unit are, which equals 10^2 square metres
> ☐ volume in litres, equal to 10^{-3} cubic metre

5.9.4   For use with units of the International System of Units, **decimal prefixes** have been increasingly elaborated, so that numerical values can almost always be between 0.1 and 999. The prefixes to terms and symbols for SI units form a series from 10^{-24} to 10^{24} (Table 5.3). The distinction between capital letters and lower-case letters is significant for the meaning of the prefix symbols, capitals being used consistently for the higher positive powers from mega (10^6).

5.9.5   For convenience, the Commission on Clinical Chemistry of IUPAC and IFCC recommended a preference in the clinical laboratory for decimal factors and decimal prefixes in **steps of a factor 1000**.[9] For most purposes, the prefixes hecto, deca, deci and centi can be avoided (Table 5.3, separated by broken lines), although they have equal legal standing.

5.9.6   A **unit together with a prefix** forms a **new unit** and is then printed according to the rules for units (§5.7). The new unit as a whole can be multiplied, divided or raised to a power.

> *Examples*
> ☐ A kilometre (km) is 1000 m.
> ☐ Square kilometre (km^2) is $(1000\,m)^2$, *i.e.* $1\,000\,000$ m,^{2} not 1000 m.^{2}

5.9.7   A **second prefix should be avoided** in units already containing a prefix, either simple units or compound units.

> *Examples*
> ☐ The substance concentration of a pollutant in air is better expressed as 8 $\mu mol\ m^{-3}$ rather than 8 $nmol\ dm^{-3}$.
> ☐ The number concentration of thrombocytes in blood should be expressed as 215×10^9 L^{-1} (not 215×10^6 mL^{-1}).

5.9.8   For historical reasons, the base unit kilogram (kg) already contains a prefix. Multiples are constructed by adding prefixes to the submultiple unit gram (g).

Examples

☐ The multiple 10^3 kg may be expressed as megagram (Mg) equal to 10^3 kg (not 10^6 kg). It is not termed kilokilogram (kkg).

☐ Multiples of the compound unit kilogram metre (kg m) should be made by replacing the k of kg rather than by inserting an additional prefix to metre: mg m, not g km.

☐ Molality should be expressed in mol kg^{-1} or mmol kg^{-1} rather than in mmol g^{-1} or μmol g^{-1} (§9.91U).

5.9.9   **A prefix should be attached to a numerator unit** rather than to a denominator unit or to a unit raised to a power, because of difficulties of interpretation. If a compound unit has to be simplified, it should first be converted to numerical factors and coherent units.

Examples

☐ $cm^{-1} = (10^{-2}$ m$)^{-1}$
$\qquad = 100$ m^{-1} (preferred)
$\qquad \neq 0.01$ m^{-1}

☐ 10^{-18} mol m^{-2} s^{-1} = amol m^{-2} s^{-1} (preferred)
$\qquad\qquad = $ mol Gm^{-2} s^{-1} = mol$(10^9$ m$)^{-2}$ s^{-1} (avoid)
$\qquad\qquad = $ mol m^{-2} Es^{-1} = mol m$^{-2}(10^{18}$ s$)^{-1}$ (avoid)

5.9.10 Special care is needed in writing **compound units** in which a letter might represent either a prefix symbol or a unit symbol.

Example

☐ N m for newton metre, rather than m N, which may be confused with millinewton, mN.

5.10   Units Outside the International System

5.10.1 Despite the general preference for coherent units (§5.8) and for expression of large or small values with prefixes (§5.9.4), CGPM[10] accepts the desire of SI users for certain units that are not part of the system and recognizes certain non-coherent units for use together with SI units (Table 5.3).

NOTE: Among 'Other non-SI units' mentioned by BIPM,[10] but not sanctioned by CGPM, are ångström, symbol Å, 1 Å $= 0.1 \times 10^{-3}$ m; bar, symbol bar, 1 bar $= 100 \times 10^3$ Pa; and millimetre of mercury, symbol mmHg, 1 mmHg $\approx 133.322\,39$ Pa.

5.10.2 A stumbling block for acceptance of the International System of Units in clinical chemistry was the **coherent unit of volume**, the **cubic metre**. The IUPAC and IFCC recommendation of 1966[9] adopted the **litre** as the preferred unit of volume in expressing concentrations, noting that calculations with a switch to the coherent unit the cubic metre required only a switch by a factor of 10^{-3} (Table 5.3). Consequently, the preferred way of expressing concentration is usually as **substance concentration** in **moles per litre** and its multiples (mol L^{-1}, mmol L^{-1}, ...) rather than in kilomoles per cubic metre and its multiples (kmol m^{-3}, mol m^{-3}, ...).

5.10.3 For **derivatives of time**, the **second** is preferred, although other units are hard to avoid.[9] **Minute, hour** and **day** are accepted for expression of time, because of their importance and widespread use. In compound units, use of minute, hour and day should be avoided to simplify comparison of data. If such stringency is not possible, the same unit of time should at least be used throughout a set of data in order to simplify comparison within one set of results.

5.11 Writing of Numbers and Expression of Numerical Values

5.11.1   In equations, tables and graphs, **numbers** should be written in upright type in Arabic numerals. The decimal mark will be a full stop (dot on the line), as is permitted by the ISO/IEC[2] in documents in English only, and will be used in this Compendium. Numbers with many digits should include spaces (but never commas or dots) between groups of three digits working out from the decimal mark. For numbers with four digits, spacing is optional. If a number is less than one, a nought should precede the decimal mark.

 Examples
- ☐ 12 345.678 901 2
- ☐ The Avogadro constant, N_A or L, equals $6.022\,141\,29(27) \times 10^{23}$ mol^{-1}; $u_r = 4.4 \times 10^{-8}$, where u_r is **relative standard uncertainty**.
- ☐ 999 hares and 1001 rabbits
- ☐ 9999 rats and 10 001 mice

5.11.2   **Numbers** may also be written in words. However, the terms for large numbers in many languages are ambiguous (Table 8.1U) and should be avoided.

 Examples
- ☐ billion may mean either 1 000 000 000 (short scale, alternatively termed **milliard**), or 1 000 000 000 000 (long scale)
- ☐ trillion may mean 10^{12} (short scale) or 10^{18} (long scale) (Table 8.1U).

5.11.3   Tabulated values of very large or very small quantities of dimension one may be labelled with the symbol for the quantity divided by the symbol for the unit or by a decimal factor and the unit. Alternatively, the heading may provide specifications about the generic quantity and indicate the unit in parenthesis. The same conventions are applicable in labelling of axes of graphs.[2]

 Example
- ☐ heading $N/10^9$ or Number of *E. coli* in 24 h urine (in 1 000 000 000), for which a tabulated value of 6 then means $N = 6\,000\,000\,000$

5.11.4   A set of values may be indicated by braces, {}, with space, or comma and space between the values.

 Example
- ☐ {2.1 2.3 2.2} mol L^{-1} or {2.1, 2.3, 2.2} mol L^{-1} for a set of three values 2.1 mol L^{-1}, 2.3 mol L^{-1} and 2.2 mol L^{-1}

5.11.5   Open and closed intervals may be indicated with square brackets enclosing the limits of the interval. The brackets are reversed for an open interval.

 Examples
- ☐ $c = [4, 8]$ mol L^{-1} means 4 mol L$^{-1} \leq c \leq 8$ mol L^{-1}
- ☐ $c =]4, 8]$ mol L^{-1} means 4 mol L$^{-1} \geq c \leq 8$ mol L^{-1}

5.12 Symbolizing Mathematical Operations

5.12.1   **Multiplication** may be indicated by a spaced multiplication sign (×) between two numbers or between symbols for two quantities, a spaced half-raised dot (·) between numbers, between symbols for two quantities or between two unit symbols, or by space between a number and a symbol for a quantity or unit. Between single-letter symbols for quantities or between a number and a symbol for a quantity, multiplication may also be indicated without space.

 In vector notation, the raised dot and the cross (×) have distinct multiplicatory functions for a **scalar** and **vector product**, respectively. If a dot on the line is used as decimal mark the raised dot should not be used as a multiplication sign between numbers.[2] An asterisk is not recognized as a multiplication sign except in some computer-programming languages.

Examples

☐ $2\ a \times b = 2\ a \cdot b = 2\ ab = 2ab$
☐ $2\ \text{kg}\ \text{s} = 2\ \text{kg} \cdot \text{s}$
☐ $(20 \pm 0.2)\ °\text{C} = 20\ °\text{C} \pm 0.2\ °\text{C}$; do not write $20 \pm 0.2\ °\text{C}$
☐ $5\ \text{N}\ \text{m} = 5\ \text{m}\ \text{N} = 5\ \text{N} \cdot \text{m}$; do not write 5 mN for 5 Nm
☐ $2 \times 10^6\ \text{kg} = 2 \cdot 10^6\ \text{kg} = 2\,000\,000\ \text{kg}$
☐ $2.3\ \text{kg} = 2{,}3\ \text{kg} \neq 2 \cdot 3\ \text{kg} = 6\ \text{kg}$

5.12.2   **Division** between two numbers, quantities or units may be indicated by a horizontal bar, an oblique stroke (/) or by a negative power without or with a mark of multiplication.

Example

☐ $\dfrac{m}{V} = m/V = mV^{-1} = m \cdot V^{-1}$

Example

☐ 5 mmol/L; 110/m;[2] 50/s

Where an expression includes more than one mark of division, brackets are necessary to avoid ambiguity. The standard ISO/IEC 80000-1[2] recommends brackets also if an expression includes a division sign followed by a multiplication sign.

Examples

☐ $(a/b)/c = a/(b\ c) = a\ b^{-1}\ c^{-1}$, not $a/b \cdot c$ nor $a/b\ c$
☐ 5 mol/m²/s is ambiguous, meaning either $5\ \text{mol}\ \text{m}^{-2}\ \text{s}^{-1}$ or $5\ \text{mol}\ \text{s}\ \text{m}^{-2}$.

5.12.3   Rules for physical notation differ from those of some computer programs.

Example

☐ In physical notation, $a + b/c - d = a + (b/c) - d$

whereas the computer notation $A + B/C - D$ often corresponds to the physical notation $[(A + B)/C] - D$

NOTE: Brackets are essential for the expression $(a + b)/(c - d)$.

5.12.4   A **mathematical operator** should be spaced from the following expression if its symbol consists of more than one character but is unspaced if it consists of a single character.[2]

Examples

☐ $\sin v$
☐ $\lg 10 = \log_{10} 10 = 1$
☐ $\Delta x = x_1 - x_0$
☐ $\dot{m} = \mathrm{d}m/\mathrm{d}t$
☐ $\int x\ \mathrm{d}t$ but $\int_0^t (x\ \mathrm{d}t)$

References

1. BIPM; IEC; IFCC; ILAC; ISO; IUPAC; IUPAP; OIML, 2012. *International vocabulary of metrology – Basic and general concepts and associated terms VIM*, 3rd edn. JCGM 200: 2012. This 3rd edition is also published as ISO Guide 99 by ISO (ISO/IEC Guide 99-12: 2007). Replaces the 2nd edition, 1993. Available on the Web site of the BIPM: http// www.bipm.org. (acc. 2015-12).
2. ISO/IEC 80000-1:2009/ Cor 1:2011. *Quantities and units — Part 1: General. Supersedes ISO 31-0: 1992 with the original title Quantities and units — Part 1: General principles.*

3. I. Mills, T. Cvitaš, K. Homann, N. Kallay, and K. Kuchitsu. *Quantities, Units and Symbols in Physical Chemistry — The IUPAC Green Book*, 3rd edn. RSC Publishing, Cambridge, 2007.

4. ISO 31:1993 (14 parts). *Quantities and units*, International Organization for Standardization, Geneva. Superseded by ISO/IEC 80000:2006-2009.

5. J. G. Hill, Terms and symbols for human body fluids and cells in laboratory medicine, *JIFCC*, 1991, **3**, 140–142.

6. IUPAC Section on Clinical Chemistry, Commission on Quantities and Units in Clinical Chemistry and IFCC Committee on Standards, Expert Panel on Quantities and Units, Approved Recommendation (1978). List of Quantities in Clinical Chemistry, *Clin. Chim. Acta*, 1979, **96**, 157F–183F.

7. IUB Nomenclature Committee, Units of Enzyme Activity, Recommendations 1978, *Eur. J. Biochem.*, 1979, **97**, 319–320.

8. D. A. Lowe. *A guide to international recommendations on names and symbols for quantities and on units of measurement*. WHO, Geneva, 1975.

9. R. Dybkær and K. Jørgensen. IUPAC–IFCC (International Union of Pure and Applied Chemistry, and International Federation of Clinical Chemistry, Commission on Clinical Chemistry). *Quantities and Units in Clinical Chemistry, including Recommendation 1966*. Munksgaard, København, 1967.

10. *Le Système International d'Unités (SI), The International System of Units (SI)*, 8th edn. (Bureau International des Poids et Mesures, Sèvres, France, 2006); it is the official French text followed by an English translation; commonly called the BIPM SI Brochure. Available on the Web site of the BIPM: www.bipm.org/en/si/ (acc. 2015-12).

11. IUPAC-IUB Joint Commission on Biochemical Nomenclature (JCBN) and Nomenclature Commission of IUB (IUB-NC), Newsletter 1981, *Eur. J. Biochem.*, 1981, **114**, 1–4; *Arch. Biochem. Biophys.*, 1981, **206**, 458–462; Hoppe-Seyler's, *Z. Physiol. Chem.*, 1981, **362**, I–IV; *J. Biol. Chem.*, 1981, **256**, 12–14.

12. P. J. Mohr, B. N. Taylor and D. B. Newell, Codata recommended values of the fundamental physical constants, *Rev. Mod. Phys.*, 2012, **84**, 1527–1605. It replaces the previously recommended 2006 CODATA set. http://physics.nist.gov/cuu/Constants/codata.pdf (acc. 2015-12).

Requesting, Generating and Transmitting Clinical Laboratory Information

6.1 General Considerations

6.1.1 To understand messages to or from a clinical laboratory, considerable local knowledge as well as specialist knowledge is necessary, because traditional terms for dedicated kinds-of-property assume such knowledge and have grown up with local needs.

Example

□ The term "magnesium" is not sufficient alongside a numerical value of substance concentration of Magnesium ions in Blood of a stated person at a stated time; it could refer to a mass concentration, the system might alternatively be Blood plasma or Erythrocytes, and the component might be either Magnesium ion or Magnesium II.

6.1.2 In order to generate systematic terms for dedicated kinds-of-property including all pertinent information in a regular format, certain rules must be followed. The scientific principles have been given in Sections 4 and 5. Rules for exchange of information within and from the clinical laboratory are given in this section. They attempt to strike a balance between clarity and brevity. Rules for architecture and organization in electronic transmission of clinical laboratory information were formulated by the European Committee for Standardization (CEN) Technical Committee 251 "Medical Informatics", based on the IFCC–IUPAC recommendation 1966.[1] A structured syntax was developed by the Committee on Nomenclature for Properties and Units (C-NPU) of IUPAC, IUB, and IFCC[2–4] and is often referred to by the term "NPU format". A user's guide has been recently published.[5–7]

6.2 Process of Requesting and Reporting

6.2.1 When laboratory data about a patient are required, the beginning of the usual chain of events is that an authorized person forwards a **request** to the clinical laboratory for the **examination of a property** related to a certain patient at a specified time and under specified conditions.

6.2.2 The formal systematic description of the property requires the choice of a **specified patient** or **part of a patient** as a system (§4.2.1), a defined **property** of which is

Compendium of Terminology and Nomenclature of Properties in Clinical Laboratory Sciences: Recommendations 2016
Edited by Georges Férard, René Dybkaer and Xavier Fuentes-Arderiu
© International Union of Pure and Applied Chemistry 2017
Published by the Royal Society of Chemistry, www.rsc.org

examined (§4.5). The system is commonly blood, blood plasma, urine, an organ or the whole patient. Especially in functional examinations an organ or the whole patient may be considered.

6.2.3   For the examination, adequate material is collected from the patient as a **primary sample**, *e.g.* blood or urine, and all or part of it is transferred to the clinical laboratory as a clinical **laboratory sample or biological sample** (§4.2.5).

6.2.4   The sample as such, or a representative part of it, in analytical chemistry termed an **analytical sample**, from which an analytical portion (§4.2.6) may be taken and used for examination (§4.2.5).
Example
□ blood taken for measurement of number concentration of erythrocytes
Sometimes the primary sample is not examined as such. It is subjected, for instance, to **fractionation** or other treatment, and a new system is given in the name of the property.
Example
□ plasma fractionated from a given primary sample of blood for measurement of substance concentration of phosphate in blood plasma
A sample of this modified material is then the **analytical sample**, from which an **analytical portion** may be taken.
After examination, a report is sent to the requesting authority.

6.3   Clinical Laboratory Report

6.3.1   Each item of data must be constructed for easy completion, reference, transmission and archiving. The International Standard ISO 15189[8] contains a subclause devoted to the format of the clinical laboratory report (*i.e.* electronic or paper) and the manner in which it is to be communicated from the laboratory.

6.3.2   According to ISO 15189,[8] essentially the report shall include, but not be limited to, the following:
□ clear, unambiguous identification of the property including, where appropriate, the examination procedure;
□ identification of the laboratory that issued the report;
□ unique identification and location of the patient, where possible, and destination of the report;
□ name or other unique identifier of the requester and the requester's address;
□ date and time of primary sample collection, when available and relevant to patient care, and time of receipt by the laboratory;
□ date and time of release of report, which, if not in the report, shall be readily accessible when needed;
□ source and system (or primary sample type);
□ results of the examination reported in SI units or units traceable to SI units, where applicable;
□ biological reference intervals, where applicable;
□ interpretation of results, where appropriate;
□ other comments (*e.g.* quality or adequacy of primary sample which may have compromised the result, results/interpretations from referral laboratories, use of any experimental procedure); the report should identify examinations undertaken as part of a development programme and for which no specific claims on examination performance are made; where applicable, information on detection limit and uncertainty of examination should be provided upon request;

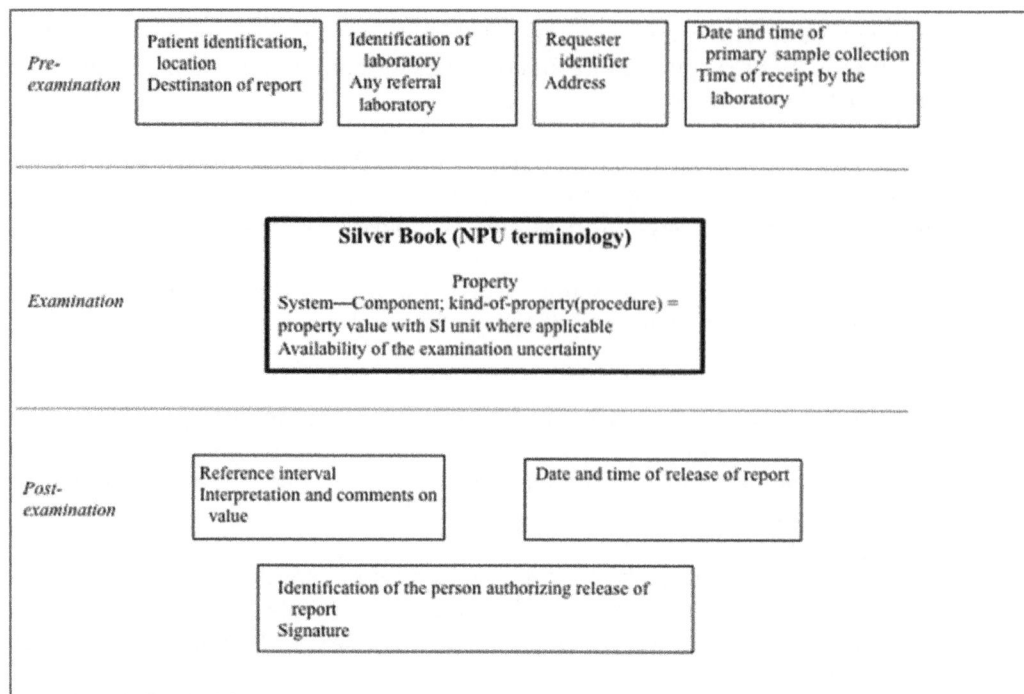

Figure 6.1 Elements of a clinical laboratory report.

    ☐ identification of the person authorizing the release of the report;

    ☐ if relevant, original and corrected results; and

    ☐ signature or authorization of the person checking or releasing the report, where possible.

This may be illustrated in a scheme (Figure 6.1) in which the different elements of a clinical laboratory report are divided into three parts corresponding to the pre-examination, examination, and post-examination phases.

6.3.3   The description of examinations performed and their results should follow the nomenclature and syntax recommended by, but not be limited to, the following organizations:

- International Bureau of Weights and Measures (BIPM);
- European Committee for Standardisation (CEN);
- International Council for Standardization in Haematology (ICSH);
- International Federation of Clinical Chemistry and Laboratory Medicine (IFCC);
- International Organization for Standardization (ISO);
- International Society of Haematology (ISH);
- International Society of Thrombosis and Haemostasis (ISTH);
- International Union of Biochemistry and Molecular Biology (IUBMB);
- International Union of Immunological Societies (IUIS);
- International Union of Microbiological Societies (IUMS);
- International Union of Pure and Applied Chemistry (IUPAC);
- Joint Committee for Guides in Metrology (JCGM); and
- World Health Organization (WHO).

Main recommendations for describing examinations and their results are compiled in the following clauses of this section.

6.4   Parts of the Systematic Name for an Examined Property

6.4.1   IUPAC, IFCC and CEN recommend the following segments of information for each examined property:

Identification of examined person; date-time

System(specification)—Component(specification); kind-of-property(specification) = $\{x\} \cdot [x]$ or $\{a\}$ or $\{a_1, a_2, a_3, \dots \}$ where $\{x\}$ stands for the numerical value of a quantity and $[x]$ for unit, together giving the measured quantity value; $\{a\}$ is for an ordinal quantity value or a nominal property value; $\{a_1, a_2, a_3, \dots \}$ is a set of ordinal or nominal property values.

The NPU format of the data remains as recommended in 1966[1] for quantities, and generalized to all properties in 1995,[9] apart from the use of a semicolon (recommended in 1991, instead of a comma in 1966) between component and kind-of-property. Supplementary information may be added (without preceding space) in parenthesis after each part on the left of the equation sign.

An em-dash (or typed double hyphen) is placed between the designations of the system and component, both of which are written with an initial capital. A semicolon is placed between the terms for the component and the kind-of-property written with an initial lower-case letter.

6.4.2   To save space, some parts of the information may be abbreviated (§6.5.1 and 6.7.2). However, because specialization in the clinical laboratory sciences precludes wide insight by individual clinical and laboratory staff into more than a few specialities, the number of abbreviations permitted must be limited and should be recognizable in a variety of linguistic areas, for example by derivation from Latin or Greek.

Examples
- Plasma(Blood)—Acetoacetate; substance concentration $= 20 \times 10^{-6}$ mol L^{-1}
  or P(B)—Acetoacetate; subst.c. $= 20$ µmol L^{-1}
- Patient—Blood; volume $= 5.2$ L
  or Pt—Blood; vol. $= 5.2$ L

6.5   Systems Investigated in Clinical Laboratory Sciences

6.5.1   Among the numerous systems in the living person that are accessible for investigation by clinical laboratories, many have received **abbreviations** in several languages as recognized by some international, national or regional organizations.[10] For systems (or components) defined as a type of entity, the distinction between singular and plural (with s) may be used to indicate whether the object of the study is a single entity or a collection of entities. Abbreviations are written with an initial capital (Table 6.1).

6.5.2   Some systems require further specification (*e.g.* arterial, arteriolar or capillary, fasting, venous) and abbreviations in several languages have been recommended.

6.5.3   Sometimes it is helpful to specify a main system and, in parenthesis, a supersystem or a subsystem (§4.2.8). The details depend on convenience and on information required.

Examples
- Erythrocytes(Blood)—Erythroblasts; number fraction $= 0.02 = 2$ %
  or with abbreviations
  Ercs(B)—Erythroblasts; num.fr. $= 0.02 = 2$ %
- Patient(Urine)—Sodium ion excretion; substance rate(24 h) $= 170$ mmol L^{-1}
  or abbreviated
  Pt(U)—Sodium ion excretion; subst.rate(24 h) $= 170$ mmol L^{-1}

Alternatively, the same physiological function may be expressed as follows:
- Urine—Sodium ion; amount-of-substance(24 h) $= 170$ mmol
  or abbreviated
  U—Sodium ion; am.s.(24 h) $= 170$ mmol

For an extensive list of English-language abbreviations for kinds-of-property, see Table 6.2.

Table 6.1 Abbreviations for systems in the human body. They were developed for the English language[10] and by the Danish data bank of dedicated kinds-of-property in clinical laboratory sciences.[11] The distinction between singular and plural (by an s) may be used to indicate whether the object of the study is a single entity or a collection of entities.

Term	Abbreviation	Term	Abbreviation
Acid mucopolysaccharides	AMPS	Lactate dehydrogenase	LDH
Alpha-1-fetoprotein	Fetoprot	Lavage fluid	LavageF
Amniotic fluid	AmF	Leukocyte(s)	Lkc(s)
Amniotic fluid cell protein	AmFCProt	Leukocyte protein	LkcProt
Antitrypsin	Atrp	Liver cell(cultured) protein	LiverC(cult)Prot
Aqueous solution	AqSolu	Lymphocyte(s)	Lymc(s)
Ascitic fluid	AscF	Lymphocytes(Blood) + Plasma	Lymcs(B) + P
Aspirate	Aspir		
Basophilocyte(s)	Basoc(s)	Lymphocyte protein	LymcProt
Blood	B	Mammary cytosol protein	MammCytosolProt
Blood fraction	Bfract	Muscle protein	MuscleProt
Blood plasma	P	Neutrophilocyte(s)	Ntrc(s)
Blood serum	S	Patient	Pt
B-lymphocyte(s)	B-lymc(s)	Pericardial fluid	PericardF
Bone marrow	Marrow	Pituitary gland	PitGl
Bursa fluid	BursaF	Plasma	P
Central nervous system	CNS	Pleural fluid	PlF
Cerebrospinal fluid	CsF	Protein	Prot
Cerebroventricular fluid	CvF	Reticulocyte(s)	Rtc(s)
Cervical mucus	CerMu	RNA, Messenger	mRNA
Choline esterase	ChE	Saliva	Sal
Chorionic villus cell protein	ChorVillCProt	Secretion	Secr
		Semen	Sem
Chorionic villus cell(cultured) protein	ChorVillC(cult)Prot	Semen(Patient identification) + Cervical mucus(Patient identification)	Sem(PtID) + CerMu(PtID)
Creatine kinase	CK		
Cyst fluid	CystF	Seminal plasma	SemP
Duodenal fluid	DuodF	Sexual-hormone-binding globulin	SHBG
Endometric cytosol protein	EndomCytosProt		
		Spermatozoa	Spmzoa
Erythrocyte(s)	Erc(s)	Stomach fluid	StomF
Expectorate	Ex	Synovial fluid	SynF
Extracellular fluid	EcF	Thrombocyte(s)	Trc(s)
Faeces	Faes	System	Syst
Fibroblast protein	FibroblProt	T-lymphocyte(s)	T-lymc(s)
Fistula fluid	FistulaF	Urine	U
Fluids and secretions	Fluid + Secr	Vaginal fluid	VagF
Haemoglobin	Hb	Vesicle fluid	VesicF
Intestine, small	Intest(small)	Wound fluid	WoundF
Kidney	Kidn		

6.5.4   If the patient is **fasting**, that information may be indicated after the term for the system, supersystem or subsystem.

Examples

☐ Blood(capillary; fasting patient)—Glucose; substance concentration = 7.0 mmol L^{-1}

or abbreviated

B(c; fPt)—Glucose; subst.c. = 7.0 mmol L^{-1}

6.6   Component

6.6.1   **General chemical** and **biochemical** components should be described by the recommended trivial (bio)chemical term (IUPAC Division VIII;[12,13] IUPAC/IUBMB JCBN;[13,14] IUPAC/IUBMB JCBN).[4]

Table 6.2 English-language abbreviations for kinds-of-property used in clinical laboratory sciences.

Term	Abbreviation
amount-of-substance	am.s.
arbitrary (modifier for specific kind-of-property)	arb.
catalytic-activity concentration	cat.c.
catalytic-activity content	cat.cont.
catalytic-activity fraction	cat.fr.
catalytic-activity rate	cat.rate
catalytic ratio	cat.ratio
concentration	conc.
diameter	diam.
dynamic viscosity	dyn.visc.
entitic amount-of-substance	entitic am.s.
entitic catalytic-activity	entitic cat.act.
entitic number	entitic num.
entitic volume	entitic vol.
entitic volume difference	entitic vol.diff.
kinematic viscosity	kin.visc.
mass concentration	mass c.
mass fraction	mass fr.
molality	molal.
number concentration	num.c.
number fraction	num.fr.
number content	num.cont.
number of entities	num.entities
partial pressure	part.pr.
pressure	pr.
reciprocal relative time	recip.rel.time
relative amount-of-substance	rel.am.s.
relative entitic number	rel.entitic num.
relative mass concentration	rel.mass c.
relative mass ratio	rel.mass ratio
relative number concentration	rel.num.c.
relative number ratio	rel.num.ratio
relative substance concentration	rel.subst.c.
relative time	rel.time
relative viscosity	rel.visc.
relative volumic mass	rel.volumic mass
saturation	sat.
saturation fraction	sat.fr.
sequence variation	seq.var.
substance concentration	subst.c.
substance content	subst.cont.
substance content rate	subst.cont.rate
substance fraction	subst.fr.
substance rate	subst.rate
substance ratio	subst.ratio
surface	surf.
temperature	temp.
temperature difference	temp.diff.
threshold mass concentration	threshold mass c.
threshold substance concentration	threshold subst.c.
time difference	time diff.
velocity	vel.
velocity ratio	vel.ratio
volume	vol.
volume content	vol.cont.
volume fraction	vol.fr.
volume rate	vol.rate
volumic number	volumic num.

6.6.1.1 **Abbreviations** have not been accepted for systematic terms for components, because they are not internationally accepted, are not interdisciplinary, are often
derived from obsolete terms (†), and may be ambiguous.

> *Examples not to be used → recommended*
> ☐ ALT → Alanine transaminase
> ☐ CK → Creatine kinase
> ☐ FSH → Follitropin
> ☐ GH: † Growth hormone → Somatotropin
> ☐ GOT: † Glutamic oxaloacetic transaminase → Aspartate transaminase
> ☐ NEFA: † Non-esterified fatty acids → Aliphatic carboxylate
> ☐ PBG → Porphobilinogen
> ☐ T_3 → Triiodothyronine
> ☐ T_4 → Thyroxin
> ☐ TSH → Thyrotropin
> ☐ UFA: Unesterified fatty acids → Aliphatic carboxylate

6.6.2   Molecular entities corresponding to **pharmaceutical substances** should be described
using the International Non-proprietary Names (INN) given by the WHO.[15]

6.6.3   Symbols of **genes** should be described according to the HUGO Gene Nomenclature
Committee (http://www.genenames.org).

6.6.4   **Bacteria** should be described by their scientific terms, which can be accessed at the
URL (http://www.bacterio.cict.fr).

6.6.5   **Fungi** should be described by their scientific terms, which can be accessed at the
URL (http://www.indexfungorum.org/Names/NAMES.ASP).

6.6.6   **Viruses** should be termed as recommended by the International Committee on
Taxonomy of Viruses. This recommendation can be accessed at the URL (http://
www.ncbi.nlm.nih.gov/ICTVdb/Ictv/fr-index.htm).

6.6.7   **Allergens** and other immunological molecules should be described according to the
International Union of Immunological Societies (IUIS) and WHO recommendations
(http://www.allergen.org/).

6.6.7.1 When the scientific term for an **animal species** related to an antigen is to be given,
the Latin terms recommended by the International Commission on Zoological
Nomenclature (http://iczn.org) should be used.

6.6.7.2 When the scientific term for a **plant species** related to an antigen is to be given, the
Latin terms recommended by the International Plant Name Index (http://www.ipni.
org) should be used.

6.6.8   **Agricultural chemicals**, including **pesticides**, should be described by the ISO
recommended terms.

6.7   Kind-of-property

6.7.1   IUPAC recommended terms and symbols for kinds-of-quantity are given in the
'Green Book',[16] and terms are shown in bold type in Sections 8 and 9.

6.7.2   As mentioned in §5.5, in contrast to practice in physics and physical chemistry, where
only symbols are used, for clinical laboratory sciences a set of **abbreviations** in many
languages has been developed for commonly used kinds-of-property, including
kinds-of-quantity, and some of the common words in their terms[10] (Table 6.2).

6.7.3   Terms for kinds-of-property may be augmented by one or more specifications in parenthesis:
- examination conditions
- examination procedure
- moment or duration of time of examination (§6.8)

Examples
- ☐ relative substance concentration(actual/reference)
- ☐ substance rate(24 h)
- ☐ P—Aspartate transaminase; cat.c.(IFCC 2002)[17]

6.7.4   Since the recommendation of 1966, ways have been devised for terming most of the dedicated kinds-of-property not commonly encountered in other areas of science. Details are given in IUPAC–CQUCC and IFCC–CQU 1995[9] and in the IUPAC–IFCC Silver Book of the same year.[18]

6.8   Time

6.8.1   In clinical laboratory sciences, the report must contain a statement of the date and often time of day at which each primary sample is taken. It may be specified in a format based on ISO 8601,[19] and given as a specification to system.

Example
- ☐ 2016-04-29T16:35

6.8.2   If a previous event is known to modify the property, it should be specified after the term for the kind-of-property:

System—Component; kind-of-property(moment of time of event or time elapsed since the event)

The format is appropriate for functional examinations:

Example
- ☐ P(B)—Glucose; subst.c.(60 min after oral intake, n(glucose) $=$ 278 mmol) $=$ 8.5 mmol L^{-1}

6.8.3   If a **mean rate** is reported, the time interval may be specified in one of three ways as parenthetic specification to the kind-of-property
- initial and final moments of time (t_0/t_1)
- initial moment and time difference $(t_0/\Delta t)$
- time difference and final moment of time $(\Delta t/t_1)$

The first two presentations are preferred.

Example
- ☐ Pt(U)—Porphyrin excretion; subst.rate(2016-04-25T09:45/28T09:45) $=$ 150 μmol d^{-1}
  or
- ☐ Pt(U)—Porphyrin excretion; subst.rate (2016-04-25/3 d) $=$ 150 μmol d^{-1}
  or
- ☐ Pt(U)—Porphyrin excretion; subst.rate(3 d/2016-04-28T09:45) $=$ 150 μmol d^{-1}

6.8.4   If short-term variation is relevant, for instance with examinations lasting part of a day, specification only of duration of examination is incomplete information, for instance because of diurnal physiological cycles, and clock-times must be stated.

Example
- ☐ Pt(U)—Ammonium excretion; substance rate(2016-04-25T16:00/24:00) $=$ 180 nmol s^{-1}
  or
- ☐ U(Pt)—Ammonium; am.s.(2016-04-25T16:00/24:00) $=$ 21 μmol, for a collection starting at 16:00 and finishing at 24:00

6.8.5   In some examinations, the time elapsed during a process is a property of the system
         and is the value transmitted by the laboratory.
         Example
         ☐ B—Coagulation, surface induced; time(proc.) = 33 s

6.9   Numerical Value and Measurement Unit

6.9.1   Numerical values should be written in accordance with international rules (§5.11).

6.9.2   SI units and their multiples are indicated according to principles given in §5.7–5.10.
         Besides those units and recognized non-SI units (§5.10), clinical laboratory sciences
         also have need of other non-SI units, including WHO international units (IU) (§10.2.3).

6.9.3   The unit must never be omitted in reporting examination results, except for the
         unit 1. Nominal and ordinal examination results have no units (Section 8).

6.10   Measurement Uncertainty

The concepts considered in this section are related to measurement of quantities having
values on rational or differential scales, but not on an ordinal scale.

6.10.1   Full interpretation of a **measured quantity value** [ref. 20, concept 2.10] requires
         evaluation of the doubt attached to the value. Manufacturers and laboratories
         must supply information that can be used in assessing the **measurement un-
         certainty** [ref. 20, concept 2.26].

6.10.2   Traditionally doubt has been expressed as (relative) **measurement bias** [ref. 20,
         concept 2.18] and measures of **measurement imprecision**, which were attributed to
         the **systematic errors** [ref. 20, concept 2.17] and the **random errors** [ref. 20, concept
         2.19] of a measurement, respectively. Appropriate statistics for random effects are
         standard deviation s, **variance** s^2, **relative standard deviation** (or *coefficient of
         variation*) s_r, commonly abbreviated *CV*. Measurement bias was often reduced, by
         use of an **addend** or a **correction** [ref. 20, concept 2.53] in the form of a factor, whose
         derivation needed to be explained. Since about 1980, the international metrological
         organizations have recommended the broader concept of measurement uncer-
         tainty. The JCGM guide,[21] and its supplements 1 and 2,[22,23] available from national
         standards institutions, should be consulted for procedures of calculation.

6.10.3   **Measurement uncertainty** is a non-negative parameter characterizing the dispersion
         of the quantity values being attributed to a measurand, based on the information
         used [ref. 20, concept 2.26]. Measurement uncertainty includes components arising
         from systematic effects, such as components associated with corrections and the
         assigned quantity values of measurement standards, as well as the **definitional
         uncertainty** [ref. 20, concept 2.27]. Sometimes estimated systematic effects are not
         corrected for but, instead, associated measurement uncertainty components are
         incorporated [ref. 20, concept 2.26, Note 1]. The parameter may be, for example, a
         standard deviation termed **standard measurement uncertainty** [ref. 20, concept 2.30]
         (or a specified multiple of it), or the half-width of an interval, having a stated
         coverage probability [ref. 20, concept 2.26, Note 2]. Measurement uncertainty com-
         prises, in general, many components. Some of these may be evaluated by **Type A
         evaluation of measurement uncertainty** [ref. 20, concept 2.28] from the statistical
         distribution of the quantity values from series of measurements and can be
         characterized by standard deviations. The other components, which may be assessed
         by **Type B evaluation of measurement uncertainty** [ref. 20, concept 2.29], can also be
         characterized by standard deviations, assessed from probability density functions
         based on experience or other information [ref. 20, concept 2.26, Note 3].

6.10.4   For a given measuring system, measurement procedure and operator, components of measurement uncertainty arise from the following:
  – definition of the measurand;
  – unrepresentative sampling of the system;
  – lack of knowledge of the influence of ambient conditions on the measurand or imperfect measurement of ambient conditions;
  – bias by the operator in reading an analogue instrument;
  – resolution, sensitivity and limit of detection of the measuring system;
  – values assigned to the calibrators and other reference materials (§6.11);
  – values of the constants, coefficients and factors obtained from external sources and used in calculation of data (in chemistry, for instance, molar masses (§9.92.1)); and
  – approximations and assumptions in the measurement procedure, including algorithms or equations for calculations and corrections.
  These sources of measurement uncertainty are not necessarily independent of one another.

6.10.5   Each component of measurement uncertainty estimated by Type A or B evaluation can be characterized by a variance, u^2, that may be calculated from the distribution of values with repeated measurements (Type A) or assessed by using available knowledge (Type B). The positive square root of such a variance is termed **standard measurement uncertainty**, u, and the positive square root of the sum of the variances and possible covariances is termed the **combined standard measurement uncertainty**, u_c [ref. 20, concept 2.31]. Such a quantity has the same dimension as the quantity being measured and is expressed in the same unit.

6.10.6   To meet some needs, for instance in health and safety, where an interval should be provided in which the measured quantity value can be expected to lie with a high level of confidence, the combined standard uncertainty may be multiplied by a **coverage factor** [ref. 20, concept 2.26], k, with a value usually between 2 (confidence level ≈ 95 %) and 3 (confidence level ≈ 99 %). The resulting parameter is termed **expanded measurement uncertainty**, U [ref. 20, concept 2.38]. The coverage factor and associated probability must always be stated.

6.10.7   The measurement uncertainty increases as one moves down the **calibration hierarchy** from a primary measurement standard, reference measurement material or calibrator (§6.11), through secondary measurement standard, reference measurement standard (reference material or calibrators), and working measurement standards, to the results of routine measurements. In principle, the **uncertainty budget** [ref. 20, concept 2.33] includes:
  – all corrections, factors and coefficients used;
  – procedures used to calculate the measured quantity value; and
  – all components of measurement uncertainty and the way they are evaluated.
  In essence, the questions must be asked:
  – has enough clear information been provided to allow the measurement result to be updated if new information becomes available? and
  – may the measurement result be used in calculations with other measurement results?

6.10.8   The metrological authorities[21] recommend that the measurement result be reported in one of the following ways (taking an example of a standard mass of 100 g, m_s). Words in braces may be omitted if the convention is stated elsewhere in the document.
  (1) $m_s = 100.021\,47$ g {with the combined standard measurement uncertainty, $u_c = 0.000\,35$ g}.

(2) $m_s = 100.021\ 47(35)$ g {where the number in parentheses is the numerical value of the combined standard measurement uncertainty, u_c, referred to the corresponding last digits of the measured quantity value}.

(3) $m_s = 100.021\ 47(0.000\ 35)$ g {where the number in parentheses is the numerical value of the combined standard measurement uncertainty, u_c, expressed in the unit of the measured quantity value}.

(4) $m_s = (100.021\ 47 \pm 0.000\ 79)$ g {where the number after the sign $\pm$ is the numerical value of an expanded measurement uncertainty, $U = ku_c$}. In this example, k (the coverage factor) $= 2.26$ for $v = 9$ degrees of freedom, the interval ($\pm$ 0.000 79 g) is estimated to have a level of confidence of 95 %.[22]

6.10.9 The measured quantity value and its measurement uncertainty should not be stated with an excessive number of digits. Usually u_c or U with the coverage factor needs to be stated to two significant digits, although a third digit may sometimes be necessary to avoid errors of rounding in subsequent calculations.

6.11 Measurement Standard, Reference Material and Calibrator

The concepts considered in this section are related to measurements of unitary quantities. Concepts related to nominal and ordinal examinations are not included (except for definition of reference material).

6.11.1 If appropriate, information should be supplied about the measurement standards, reference material and calibrators used in the measurement, so that values as reported can ultimately be traced back to a national or international standard, and usually to an SI unit. A measurement standard is the realization of the definition of a given quantity, with stated quantity value and associated measurement uncertainty (§6.10.3), used as a reference [ref. 20, concept 5.1].

6.11.2 **Metrological traceability** is the property of a measurement result whereby the result can be related to a reference through a documented unbroken chain of calibrations (§6.11.5), each contributing to the measurement uncertainty (§6.10.3) [ref. 20, concept 2.41].

6.11.3 A **reference material** is a material, sufficiently homogeneous and stable with reference to specified properties, which has been established to be fit for its intended use in measurement or in examination of nominal properties [ref. 20, concept 5.13]. Note that the same reference material cannot be used both for calibration and quality assurance of a given measurement situation.

6.11.4 A **certified reference material** is a reference material, accompanied by documentation issued by an authoritative body and providing one or more specified property values with associated uncertainties (§6.10.3) and traceabilities (§6.11.2) using valid procedures [ref. 20, concept 5.14].

6.11.5 **Calibration** is the operation that, under specified conditions, in a first step, establishes a relation between the quantity values with measurement uncertainties (§6.10.3) provided by measurement standards and corresponding indications with associated measurement uncertainties and, in a second step, uses this information to establish a relation for obtaining a measurement result from an indication [ref. 20, concept 2.39]. A **calibrator** is a measurement standard used in calibration [ref. 10, concept 5.12].

6.11.6 A handbook of recommended reference materials for physicochemical quantities has been published by IUPAC.[24] Lists of international standards and biological reference materials are published by WHO.[25] ISO has listed the requirements for certified reference materials and the content of supporting documentation.[26]

6.12 Systems for Coding Properties in Clinical Laboratories

Several national and international systems for coding properties in clinical laboratories have been developed and used in the clinical laboratories for reporting laboratory results.

Two systems are in wider international use today; the IFCC-IUPAC NPU terminology[1] and the Logical Observation Identifiers Names and Codes (LOINC).[27]

The NPU terminology was developed by the Committee on Nomenclature for Properties and Units (C-SC-NPU) of IFCC and IUPAC. It is officially recommended by these international bodies and user's guides are available.[5-7] It adheres strictly to international terminology and metrology standards (see §6.3.3) and has had a normative impact on several later coding systems. It is in national use in some European countries and is followed in this Compendium.

The LOINC system is developed and managed by the Regenstrief Institute, an organization affiliated to the Indiana University School of Medicine in Indianapolis, Indiana. It is widely used in English-speaking areas.

Some examples are presented below for comparative purposes:

NPU   U—Creatininium; subst.c. = ? mmol L^{-1}
LOINC   Creatinine:SCnc:Pt:Urine:Qn
NPU   P—Human immunodeficiency virus 1(ag; p24); arb.c.(proc.) = ?
LOINC   HIV 1 p24 Ag:ACnc:Pt:Ser:Ord
NPU   P—Cholesterol + ester/Cholesterol + ester, in HDL; subst.ratio = ?
LOINC   Cholesterol.total/Cholesterol.in HDL:ScRto:Pt:Ser/Plas:Qn

References

1. R. Dybkær and K. Jørgensen, IUPAC–IFCC (International Union of Pure and Applied Chemistry, and International Federation of Clinical Chemistry, Commission on Clinical Chemistry). *Quantities and Units in Clinical Chemistry, including Recommendation 1966.* Munksgaard, København, 1967.

2. IUPAC Section on Clinical Chemistry, Commission on Quantities and Units in Clinical Chemistry, IFCC Committee on Standards, Expert Panel on Quantities and Units. Approved Recommendation (1978). List of Quantities in Clinical Chemistry. *Clin. Chim. Acta*, 1979, **96**, 157F–183F.

3. IUB Nomenclature Committee. Units of Enzyme Activity, Recommendations 1978. *Eur. J. Biochem.*, 1979, **97**, 319–320.

4. IUPAC-IUB Joint Commission on Biochemical Nomenclature (JCBN), and Nomenclature Commission of IUB (IUB-NC), Newsletter 1981, *Eur. J. Biochem.*, 1981, **114**, 1–4; *Arch. Biochem. Biophys.*, 1981, **206**, 458–462; Hoppe-Seyler's, *Z. Physiol. Chem.*, 1981, **362**, I–IV; *J. Biol. Chem.*, 1981, **256**, 12–14.

5. U. Magdal Petersen, R. Dybkær and H. Olesen, Properties and units in the clinical laboratory sciences. (Technical Report). Part XXIII. The NPU terminology principles, and implementation: A user's guide. *Pure Appl. Chem.*, 2012, **84**(1), 137–165. Also *Clin. Chem. Lab. Med.*, 2012, **50**, 35–50.

6. G. Férard and R. Dybkaer, Recommendations for clinical laboratory science reports regarding properties, units, and symbols: the NPU format, *Clin. Chem. Lab. Med.*, 2013, **51**(5), 959–966.

7. G. Férard and R. Dybkaer, The NPU format for clinical laboratory science reports regarding properties, units, and symbols (IUPAC Technical Report), *Pure Appl. Chem.*, 2014, **86**(12), 1923–1930.

8. ISO 15189:2007. *Medical laboratories — Particular requirements for quality and competence.* Replaced by ISO 15189:2012.

9. H. Olesen and IUPAC–IFCC (International Union of Pure and Applied Chemistry, Clinical Chemistry Division, Commission on Quantities and Units in Clinical Chemistry; and International Federation of Clinical Chemistry, Scientific Division,

Committee on Quantities and Units). Properties and units in the clinical laboratory sciences. IUPAC-IFCC Recommendations, Part I. Syntax and semantic rules, *Pure Appl. Chem.*, 1995, **67**(8/9), 1563–1574; *Eur. J. Clin. Chem. Clin. Biochem.*, 1995, **33**, 627–636; *Clin. Chim. Acta*, 1996, **245**, S5–S21.

10. J. G. Hill, Terms and symbols for human body fluids and cells in laboratory medicine, *J. Int Fed. Clin.*, 1991, **3**, 140–142.

11. http://www.labterm.dk/Enterprise%20Portal/NPU_download.aspx (acc. 2015-12).

12. N. G. Connelly, T. Damhus, R. M. Hartshorn and A. T. Hutton, *Nomenclature of Inorganic Chemistry, IUPAC Recommendations 2005, known as the IUPAC Red Book*, RSC Publishing, Cambridge, 2005.

13. H. A. Favre and W. H. Powell, *Nomenclature of Organic Chemistry (IUPAC Recommendations and Preferred IUPAC Names 2013)*, Royal Society of Chemistry, 2013.

14. *Enzyme Nomenclature 1992* [Academic Press, San Diego, California] with Supplement 1 (1993), Supplement 2 (1994), Supplement 3 (1995), Supplement 4 (1997) and Supplement 5 (in *Eur. J. Biochem.*, 1994, **223**, 1–5; *Eur. J. Biochem.*, 1995, **232**, 1–6; *Eur. J. Biochem.*, 1996, **237**, 1–5; *Eur. J. Biochem.*, 1997, **250**, 1–6 and; *Eur. J. Biochem.*, 1999, **264**, 610–650, respectively), and a www version prepared by G. P. Moss: www.chem.qmul.ac.uk/iubmb/enzyme (acc. 2015-12).

15. *International Nonproprietary Names (INN) for Pharmaceutical Substances. Names for Radicals and Groups and Others: Comprehensive List, 2012.* WHO, Geneva, 2012.

16. I. Mills, T. Cvitaš, K. Homann, N. Kallay and K. Kuchitsu, *Quantities, Units and Symbols in Physical Chemistry — The IUPAC Green Book*, RSC Publishing, Cambridge, 3rd edn, 2007.

17. G. Schumann, R. Bonora, F. Ceriotti, G. Férard, C. A. Ferrero, P. F. Franck, F. J. Gella, W. Hoelzel, P. J. Jørgensen, T. Kanno, A. Kessner, R. Klauke, N. Kristiansen, J. M. Lessinger, T. P. Linsinger, H. Misaki, M. Panteghini, J. Pauwels, F. Schiele, H. G. Schimmel, G. Weidemann and L. Siekmann, International Federation of Clinical Chemistry and Laboratory Medicine. IFCC primary reference procedures for the measurement of catalytic activity concentrations of enzymes at 37 degrees C. International Federation of Clinical Chemistry and Laboratory Medicine. Part 5. Reference procedure for the measurement of catalytic concentration of aspartate aminotransferase, *Clin. Chem. Lab. Med.*, 2002, **40**(7), 725–733.

18. J. C. Rigg, S. S. Brown, R. Dybkær and H. Olesen, Compendium of Terminology and Nomenclature of Properties in Clinical Laboratory Sciences, *The Silver Book. Recommendations 1995. IUPAC-IFCC*, Blackwell, Oxford, 1995.

19. ISO 8601:2004. *Data elements and interchange formats — Information interchange – Representation of dates and times.*

20. BIPM; IEC; IFCC; ILAC; ISO; IUPAC; IUPAP; OIML, 2012. *International vocabulary of metrology – Basic and general concepts and associated terms VIM*, 3rd edn. JCGM 200: 2012. This 3rd edition is also published as ISO Guide 99 by ISO (ISO/IEC Guide 99-12: 2007). Replaces the 2nd edn, 1993. Available on the Web site of the BIPM: http//www.bipm.org. (acc. 2015-12).

21. JCGM Guide 98-3:2008. *Uncertainty of measurement — Part 3: Guide to the expression of uncertainty in measurement (GUM:1995*. Available at bipm.org (acc. 2015-12).

22. *Evaluation of measurement data — Supplement 1 to the "Guide to the expression of uncertainty in measurement" — Propagation of distributions using a Monte Carlo method.* JCGM 101: 2008.

23. *Evaluation of measurement data — Supplement 2 to the "Guide to the expression of uncertainty in measurement" — Extension to any number of output quantities.* JCGM 102: 2011.

24. K. N. Marsh, IUPAC CPMS, *Recommended Reference Materials for the Realization of Physicochemical Properties*, Blackwell, Oxford, 1987.

25. *Biological substances: International standards and reference reagents. Reference preparations established by the Expert Committee on Biological standardization*, WHO, Geneva, 2010. http://www.who.int/biologicals.

26. ISO 15194:2009. *In vitro diagnostic medical devices — Measurement of quantities in samples of biological origin – Requirements for certified reference materials and the content of supporting documentation.*

27. A. W. Forrey, C. J. McDonald, G. DeMoor, S. M. Huff, D. Leavelle, D. Leland, T. Fiers, L. Charles, F. Stalling and A. Tullis, The logical observation identifier names and codes (LOINC) database: A public use set of codes and names for electronic reporting of clinical laboratory results, *Clin. Chem.*, 1996, **42**, 81–90.

SECTION 7

Choice and Use of Kinds-of-property for Different Examination Purposes

7.1 Expression of Composition: Amount-of-substance; Substance Concentration; Molality

7.1.1   The idea of **amount-of-substance** as a base kind-of-quantity crystallized in the 1960s and was approved by IUPAC in 1965. It replaced the less clearly defined concepts of *equivalence* and *normality*, which had evolved since the 19th century. On the advice of IUPAC, IUPAP and ISO, amount-of-substance, with the measurement unit mole, were added to the International System of Units as a seventh base kind-of-quantity and unit by the 14th General Conference on Weights and Measures (CGPM) in 1971. The term **amount-of-substance** is analogous to **amount-of-electricity, amount-of-light, amount-of-heat** or **amount-of-energy**, in which **amount** (or *quantity*) indicates an extensive kind-of-quantity (§5.4.2). Clinical laboratory sciences such as clinical toxicology, biological haematology and biological chemistry should find advantages using amount-of-substance instead of mass when the elementary entity is known.

7.1.2   In clinical laboratory sciences, the usefulness of quantities with values proportional to **number of entities** of any chemical component was soon realized. It is the number of entities in a given volume as a measure of concentration of component, neither mass nor volume of that component that influences the rate of **chemical reactions,** including **enzymic reactions** in living organisms. Moreover, the number of entities of a component also influences **biological effects** based on biochemical processes and so is applicable in descriptions of the state of a person, although its values are usually inconveniently large. **Amount-of-substance** is proportional to the number of entities of a chemical component and gives values in a more convenient numerical interval than number of entities. Both amount-of-substance and number of entities are considered as **molecular kinds-of-quantity** (or **kinds-of-quantity of a molecular nature**). **Substance concentration** (§9.88.1) is in general the most convenient kind-of-quantity of this type for many sciences; **number concentration** (§9.6.2) gives values almost 24 magnitudes larger. **Mass** (§9.8.1) and **mass concentration** (§9.14.2) are less meaningful for the interpretation of chemical reactions and biological effects of components.

   The report of the Director-General of the World Health Organization submitted to the 30th World Health Assembly in 1977[1] explained the issues as follows in its Section 2.

Compendium of Terminology and Nomenclature of Properties in Clinical Laboratory Sciences:
Recommendations 2016
Edited by Georges Férard, René Dybkaer and Xavier Fuentes-Arderiu
© International Union of Pure and Applied Chemistry 2017
Published by the Royal Society of Chemistry, www.rsc.org

"When chemical substances interact, either *in vitro* or *in vivo*, the proportions in which they do so are related to their relative molecular mass ('molecular weight') [and to molar mass, §9.92.1]. This is measured in terms of 'amount of substance' by means of the mole. For a proper understanding of chemical reactions, therefore, whether they occur in the laboratory or in the body, the use of the mole is essential. The use of mass related units (such as milligrams per litre) serves no purpose other than the purely arbitrary one of deciding whether or not a given value is greater or less than a certain reference value. The expression of concentrations of substances [*i.e.* components] in body fluids in molecular terms also serves this purpose, but in addition gives valuable insight into the balance of the constituents [*i.e.* components]. Such insight cannot be obtained from [values expressed in] mass related units."

7.1.3   Molecular kinds-of-quantity reflect simple relationships through the laws of chemistry: the **law of constant proportions**, the **law of multiple proportions**, and the **law of combining proportions**. Likewise, they reflect **functional relationships in biology**.

   Examples of functionally related chemical components (termed according to trivial usage, not systematically):
   - ☐ acetoacetate – acetone – β-hydroxybutyrate
   - ☐ adrenaline – noradrenaline – 4-hydroxy-3-methoxymandelate
   - ☐ bilirubin – bilirubin monoglucuronide – bilirubin diglucuronide – bilirubin-albumin – albumin
   - ☐ calcium ion – phosphate ions
   - ☐ calcium – albumin
   - ☐ carnitine free – esterified carnitine forms – carnitine total
   - ☐ cholesterol – cholesterol ester
   - ☐ copper ions – ferroxidase
   - ☐ creatine – creatininium
   - ☐ cystine – cysteine
   - ☐ glucose – glycated haemoglobin
   - ☐ glycerol – triglyceride – glycerol total
   - ☐ haemoglobin(Fe) – dioxygen(O_2) – iron(II) – transferrin iron(III) – glucose – CO_2
   - ☐ haptoglobin – haemoglobin
   - ☐ lactate – pyruvate
   - ☐ reduced glutathione – oxidized glutathione
   - ☐ sodium ion – potassium ion
   - ☐ triiodothyronine – thyroxine

7.1.4   The physicochemical description of a system in terms of its components requires the definition of intensive kinds-of-quantity in terms of extensive kinds-of-quantity. Composition is usually stated in terms of intensive **compositional kinds-of-quantity** (Table 7.1). Conversion between different compositional kinds-of-quantity is possible by means of **material kinds-of-quantity** (Table 7.2). **Molality**, n_B/m_A (§9.91.2), falls outside this scheme, but is also widely used in clinical chemistry, having advantages over substance concentration in its independence of temperature and pressure.

7.1.5   The amount-of-substance of a portion of a pure chemical component, n_{B*}, *i.e.* a defined portion of chemical substance, is proportional to its mass. The coefficient of proportionality is the **molar mass** of that substance, which may be calculated if the **molecular formula** and the **molar mass** (§9.92.1) of each constituent element are known. Values are regularly revised and published by the Committee on Data for Science and Technology.[2] Even if the molecular formula of a substance is unknown, its molar mass may be estimated empirically.

Table 7.1 Terms and symbols of compositional kinds-of-quantity, derived from two extensive kinds-of-quantity Q and Q': number, volume, mass, amount-of-substance, catalytic activity and radioactivity, $q_B = Q_B/Q'_1$.[a] For the meaning of modifier symbols, see §4.3.1.

Numerator kind-of-quantity Q_B	Denominator kind-of-quantity Q'_1			
	number $\sum_A^N N$	volume V_1	mass m_1	amount-of-substance $\sum_A^N n$
number N_B	number fraction δ_B	number concentration C_B	number content N_B/m_1	
volume V_B		volume fraction ϕ_B	volume content ν_B	
mass m_B		mass concentration ρ_B	mass fraction w_B	
amount-of-substance n_B		substance concentration c_B	substance content ν_B	substance fraction x_B
catalytic activity z_E		catalytic-activity concentration b_E	catalytic-activity content z_E/m_1	
absorbed dose A_B		absorbed dose concentration A_B/V_1	absorbed dose content a_B	

[a]⟨19⟩ The term "radioactivity" is not recommended in the 9th SI brochure draft of CCU and is replaced here by "absorbed dose".

Table 7.2 Terms and symbols of material kinds-of-quantity (fundamental constants, material constants or material coefficients) derived from two extensive kinds-of-quantity Q and Q': number, volume, mass and amount-of-substance, $q_B = Q_B/Q'_B$ or $q_1 = Q_1/Q'_1$. They can be used as coefficients of proportionality to convert compositional kinds-of-quantity listed in Table 7.1.[a] For the meaning of modifiers, see §4.3.14.

Numerator kind-of-quantity Q_B or Q_1	Denominator kind-of-quantity Q'_B or Q'_1			
	number N	volume V	mass m	amount-of-substance N
number N		volumic number N_B/V_B	massic number N_B/m_B	Avogadro constant N_A, L
volume V	entitic volume V_B/N_B		massic volume ν_1	molar volume $V_{m,B}$
mass m	entitic mass m_B/N_B	volumic mass ρ_1		molar mass $M_B, m_{m,B}$
amount-of-substance n	entitic substance constant N_A^{-1}, L^{-1}	volumic substance ν_1^{-1}	massic substance $M_B^{-1}, m_{m,B}^{-1}$	
catalytic activity z_E	entitic catalytic activity z_E/N_E			molar catalytic activity $z_{m,E}$
absorbed dose[a] A	entitic absorbed dose τ_B	volumic absorbed dose A_1/V_1	massic absorbed dose a_1	molar absorbed dose $A_{m,B}$

[a]⟨19⟩ The term "radioactivity" is not recommended in the 9th SI brochure draft of CCU and is replaced here by "absorbed dose".

7.1.6   Ionic strength of solution 1 is usually defined as half of the sum of the products of the charge number squared of each solute component, z_B^2, and a compositional kind-of-quantity of each component, $q(B^z)$:

$$I_1 = 0.5 \sum_{B}^{N} (z_B^2 q_B) \quad \text{for molecules of all ionic components B, } \ldots , N$$

$$= 0.5 \sum_{B}^{N} |z_B| \cdot q(z_B^{-1}B) \quad \text{for ionic charge entities of each ionic component}$$

The compositional kind-of-quantity taken may, for instance, be substance concentration, substance content or substance fraction (Table 7.1), but is usually molality. The concept can therefore also be considered as, for instance, the substance concentration of the half of ionic charge entities of all ionic solute components.

7.2 Thermodynamic Kinds-of-quantity: Free Components and Chemical Activity

7.2.1   The energy of a chemical component may be described in terms of its **chemical potential**. For some purposes where energy relationships are relevant, **relative chemical activity** (§8.10.2), **pH** (§8.9.4) and some other thermodynamic kinds-of-quantity may be used. Recommendations on that subject were published in 1984.[3] They are relevant to clinical laboratory scientists using **ion-selective electrodes** or some other physicochemical techniques.

7.2.2   The number of kinds-of-quantity in this area is very large (*e.g.* §9.88), partly because of a certain redundancy between different scales for definition of kinds-of-quantity of solute components, based, for instance, on substance fraction, substance concentration or molality. For the clinical laboratory, the number of kinds-of-quantity should be limited as far as possible, selecting those that are most meaningful.

7.2.3   If a component B with a known amount-of-substance is added to a closed system, it is the **stoichiometric amount-of-substance** of component, $n_{o,B}$, that is known. If the volume of the system is also known, these extensive kinds-of-quantity can be used to calculate the intensive kind-of-quantity **stoichiometric substance concentration** of component, $c_{o,B}$:

$$c_{o,B} = n_{o,B}/V_1$$

7.2.4   If the component added to the system dissociates or reacts with other components to form a series of products C, D, $\ldots$, only part of the original component B remains free in the system. So it is essential to distinguish between **stoichiometric component** and **free component**. Symbols for kinds-of-quantity with the stoichiometric component may therefore be distinguished by the modifier $_o$. In clinical chemistry, the term stoichiometric component is rarely used. Instead, the term or symbol of the component is modified to indicate the inclusion of the various products of reaction (§4.3). The modifier may alternatively be attached to the term or symbol for the kind-of-quantity, for instance, stoichiometric amount-of-substance, $n_{o,B}$, or stoichiometric substance concentration, $c_{o,B}$.

7.2.5   The **chemical potential** of component B, μ_B, is the differential change in thermodynamic energy, dU_B, divided by the differential change in the stoichiometric amount-of-substance of component B, $dn_{o,B}$, when other independent extensive kinds-of-quantity (*i.e.* volume, entropy, electrical charge, amount-of-substance of solvent, and of other stoichiometric components) are kept constant:

$$\mu_B = (dU_B/dn_{o,B})_{V,S,Q,n(A),n(C), \ldots ,n(N)}$$

7.2.6   Chemical potential is generally converted into an exponential function, chemical activity, λ_B (§8.10.2):

$$\lambda_B = \exp [\mu_B/(R \times T)]$$

7.2.7 **Chemical activity**, λ_B, cannot be measured as such, only as **relative chemical activity**, a_B (§8.10.2), relative to a standard state. For instance, **relative molal activity**, $a_{b,B}$ (§8.10.2.2), relative to the chemical activity, $\lambda_{B,\ominus}$, where the reference value of active molality of the component, $\tilde{b}_{B,\ominus}$ (equal to $\gamma_{B,\ominus}\, b_{B,\ominus}$) is 1 mol kg^{-1}:

$$a_{b,B} = \lambda_B / \lambda_{B,\ominus}$$

7.2.8 The relative molal activity may be divided by the **molal activity factor**, $\gamma_{b,B}$ (§8.10.3.2), to calculate the molality of free component:

$$b_B = (a_{b,B} / \gamma_{b,B}) \times \tilde{b}_{B,\ominus}$$

7.2.9 **Substance concentration of free component** in an aqueous solution 1 can be calculated by multiplying molality by the mass concentration of water in solution 1:

$$c_B = b_B \times \rho(H_2O)$$

7.2.10 General practice in clinical laboratory sciences is to report **substance concentration of free component** rather than chemical activity. Treatment of the following components is exceptional:
  ☐ **hydrogen ion**, reported in terms of **pH** (§8.9.4)
  ☐ **dissolved gases** (O_2 and CO_2), reported in terms of **partial pressure** (or **gas tension**) in the liquid phase, $p_{B(1)}$ (§9.41.2)
  ☐ **water**, reported in terms of **osmolality**, $\hat{b}$ (§9.91.2.3), or **osmotic concentration**, $\hat{c}$ (§9.88.3.1).

7.3 Optical Spectroscopy

7.3.1 Two commissions of IUPAC have published many recommendations on optical spectroscopy: the Commission on Molecular Structure and Spectroscopy (IUPAC-CMSS) and the Commission on Spectrochemical and other Optical Procedures for Analysis (IUPAC-CSOPA). IUPAC-CQUCC and IFCC-EPQU have published two recommendations,[4,5] from which general matters are summarized here and details of individual kinds-of-quantity are listed in Sections 8 and 9. The IFCC Expert Panel on Instrumentation has published provisional guidelines for listing procedural specifications of flame emission spectrometers and atomic absorption spectrometers.[6,7]

7.3.2 **Spectroscopy** is the study of physical systems by their interactions with electromagnetic radiation or by the electromagnetic radiation such systems emit. Spectrometry is the measurement of such radiation as a means of obtaining information about systems and their components.

7.3.3 In some types of optical spectroscopy, the radiation originates from an external source and is modified by the system under investigation. If the modification is only absorption, the study is termed **absorption spectroscopy** and the type of measurement is termed **absorption spectrometry**. Other processes observed or measured include reflection, refraction, polarization, scattering (or diffusion) and luminescence, either phosphorescence or fluorescence. The processes are indicated by subscripts to symbols:

– **reference radiation**, either incident or a zero reference	0
– **absorption**, absorbed	abs, a
– **scatter** (or **diffusion**), scattered, diffused	s, dif, d
– **reflection**, reflected	refl, r
– **transmission**, transmitted	tr, t

7.3.4 The reduction in radiation passing through a system is termed **attenuation** (or *extinction*). If **boundary effects** (*e.g.* reflection and scatter at surfaces of the cuvette) are excluded or allowed for, the resulting kinds-of-quantity are termed **internal** and their symbols may be modified by a subscript i. The reduction is normally due to both **scatter within the system** (also termed **diffusion**) and by **absorption**.

7.3.5 Special terms for many of the kinds-of-quantity exist where absorption can be assumed to be the only cause of the reduction in radiation. In the absence of scatter and luminescence, incident energy equals the sum of reflected, absorbed and transmitted energy:

$$Q_{e,0} = Q_{e,refl} + Q_{e,a} + Q_{e,t}$$

In absorption spectroscopy, the measuring system and measurement procedure is usually such that reflection is also excluded and the kinds-of-quantity can be termed internal. Absorbed energy then equals incident energy minus transmitted energy:

$$Q_{e,a,i} = Q_{e,0} - Q_{e,t}$$

These relationships apply also if absorption and transmission are expressed as energy fractions (§8.6.1), energy rate (or power) (§9.48.1) or areic energy rate (§9.45.1).

7.3.6 A measurement of the reduction in a collimated beam of a radiation can be expressed as the **energy fraction of the radiation transmitted** (or **internal transmittance** or *internal transmission factor*), τ_i or **energy fraction of the radiation attenuated**, including that **absorbed** (termed **internal absorptance** or *internal absorption factor*), α_i (§8.9.1). More commonly, the reduction is expressed as a logarithm, which may be on either a napierian scale (common in physics and engineering) or on a decadic scale (common in analytical and clinical chemistry). The decadic type is termed **attenuance** (or *internal optical density*), D_i (§8.9.2) or, if the reduction is due entirely to absorption, **decadic absorbance**, A (§8.9.3). Attenuation coefficients or absorption coefficients, derived by dividing attenuance or absorbance by path length, l, for a defined radiation and by substance concentration, c_B, are characteristic for some components and can be used to calculate substance concentration in the system under study by means of **Beer's simplification of the Lambert–Bouguer law**, which may be expressed in one of the following ways.

The energy rate, P or Φ_e, of a collimated beam of monochromatic radiation decreases logarithmically in a homogeneous isotropic medium with path length and with substance concentration or, in the gas phase, with partial pressure, p_B, of an absorbing component B:

$$P = P_0 \exp\left(-\kappa_B\, c_B\, l\right)$$

$$= P_0\, e^{-\kappa c l}$$

$$= P_0\, 10^{-\varepsilon c l}$$

where κ_B is **molar area napierian absorbance** (or **molar napierian absorption coefficient**) (§9.86.3.1) and ε is **molar area decadic absorbance** (or **molar decadic absorption coefficient**) (§9.86.3.2). The law can be generalized to other causes of reduction than absorption.

7.3.7 In other types of spectroscopy and spectrometry, energy enters the system under study, for instance electrical energy, heat (**thermal excitation**) or energy from a chemical reaction (**chemiluminescence**), and one or more components resonate between a higher and lower energy state, releasing the energy again by emission of a radiation from the system.

7.3.8   In **atomic emission spectroscopy** and in **atomic absorption spectroscopy**, the original material system is converted into an **aerosol**, usually a mist of drops, by **nebulization**. The aerosol is heated, commonly in a flame or in an electrothermal atomizer. First, any solvent evaporates, resulting in **desolvation**, the conversion of the solute to a dry aerosol. Some components of the system may be broken down by **pyrolysis**. Then the heat causes **volatilization** of the solute, its conversion to a molecular vapour. The temperature must be sufficient to cause partial or total **atomization**, the dissociation of molecules into radicals or neutral atoms in a **physicochemical plasma**, in which some components, for instance alkaline earth elements, ionize. **Ionization** is often undesirable and may be minimized by reducing the temperature of the flame or by adding electron donors, for instance caesium(II).

7.3.9   In **emission spectrometry**, energy is transferred to the molecules, radicals, atoms or ions, causing excitation and resulting in **emission bands** or **emission lines** specific for particular particles of a given component. The energy rate of a band or line (§9.46.2) depends critically on the physicochemical properties of the flame, for instance temperature, composition of the gas mixture and redox properties. For a given flame, the energy rate depends on substance concentration of the component, both in the material analysed and in the physicochemical plasma. The calibration function can be established with a series of **calibrators** (§6.11.5) with the component at known concentrations. New developments such as inductively coupled plasma (ICP) optical emission spectrometry, or ICP mass spectrometry, are used for the determination of trace elements.

7.3.10 In **atomic absorption spectrometry**, a background source, usually a hollow cathode lamp, sends a radiation through the physicochemical plasma. The radiation is absorbed (and energy is reduced) only at certain **resonance lines** of atoms from the component considered, defined by spectral position (§7.3.12–13), for instance λ_0. The energy may be released again by a rise in temperature of the flame or other physicochemical plasma. At thermodynamic temperatures below 3300 K, only a few resonance lines can be used, because few electrons occupy the higher energy levels.
    Example
    □ For Mg(II), the cathode tube will emit all the spectral lines of Mg(II) but radiation is absorbed only at $\lambda_0 = 285.2$ nm.
    Absorbance at λ_0 is proportional to substance concentration of the element in the physicochemical plasma.

7.3.11 In some atomic absorption spectrometers, a **calibration component** (or *internal standard*) with a distinct resonance line or band different from that of the component of interest can be added to the prepared material in order to allow electronic compensation for instrumental drift or variations resulting from differences of the calibration components from the material under investigation, for instance in surface tension, volumic mass (or mass density) and viscosity. Measurement of substance concentration is based on the energy ratio of the two absorption lines or bands, whereas other methods are based on the energy rate of the one line or band being directly proportional to concentration of the absorbing component.

7.3.12 In either emission or absorption spectrometry, the radiation may be considered as a physical component (§4.3.2) and its interaction with the chemical component may be specified as a process (§4.4). A given radiation may be specified:
    – by its **energy**, Q_e (§9.42.1.7), or a quantity derived from energy, such as energy rate, Φ_e or $P = dQ_e/dt$ (§9.48.1.2), or areic energy rate, $\phi_e = (dQ_e/dt)/A$ (§9.45.1.2-3);
    – by its **direction**, ignored in absorption spectrometry by studying only radiation in the direction of a collimated beam or a laser beam, but used for study of scatter with components of different direction;

Table 7.3   Terms of regions of the spectrum of electromagnetic radiation in terms of wavelength in a vacuum, λ_0, wavenumber in vacuum, $\tilde{v}$, frequency, v, entitic energy, Q_e/N_Φ, and molar energy, Q_e/n_Φ; h, Planck constant; N_A, Avogadro constant; n, amount-of-substance; N, number of entities; subscript Φ, photons; Q_e, energy of the radiation. The boundaries are arbitrary and are differently defined by different authorities, including ir., infrared; uv, ultraviolet.

Term of region	λ_0	$\tilde{v}$	v	$Q_e/N_\Phi = hv$	$Q_e/n_\Phi = N_A hv$
γ-rays	10 pm	10^7 m^{-1} 10^9 cm^{-1}	30 EHz	19.9 fJ	12.0 GJ mol^{-1}
X-rays	10 nm	10^4 m^{-1} 10^6 cm^{-1}	30 PHz	19.9 aJ	12.0 MJ mol^{-1}
vacuum uv.	200 nm	500 m^{-1} 50×10^3 cm^{-1}	1.5 PHz	993 zJ	589 kJ mol^{-1}
near uv.	380 nm	263 m^{-1} 26.3×10^3 cm^{-1}	789 THz	523 zJ	315 kJ mol^{-1}
visible	780 nm	128 m^{-1} 12.8×10^3 cm^{-1}	384 THz	255 zJ	153 kJ mol^{-1}
near ir.	2.5 μm	40 m^{-1} 4.00×10^3 cm^{-1}	120 THz	79.5 zJ	47.9 kJ mol^{-1}
mid ir.	50 μm	2 m^{-1} 200 cm^{-1}	6.00 THz	3.98 zJ	2.40 kJ mol^{-1}
far ir.	1 mm	0.1 m^{-1} 10 cm^{-1}	300 GHz	199 yJ	120 J mol^{-1}
microwaves	0.1 m	0.001 m^{-1} 0.1 cm^{-1}	3.00 GHz	1.99 yJ	1.20 J mol^{-1}
radio waves					

- by a quantity indicating **spectral position**, commonly wavelength in vacuum, λ_0 (§9.1.3.1), or wavenumber in vacuum, $\tilde{v}$ (§9.2.1.2), which varies with the medium and which is usually reported for a vacuum rather than for air (in which values are close to those in a vacuum) or for the system irradiated; alternatively spectral position can be reported in terms of frequency, v (§9.18.1.2), photon energy or molar energy (§9.103.1), which are independent of medium. The commonly used names of regions of the spectrum are not exactly delineated, apart from 'visible rays' (Table 7.3).

7.3.13   When radiation is plotted as a function of spectral position, the plot is normally a **differential quotient** of energy (or of energy rate, areic energy rate, ...) to the quantity indicating spectral position, often termed *spectral concentration* or *spectral density* of that quantity. Such a quotient is signified by a subscript for the quantity indicating spectral position. Thus for a quantity Q in a small interval $\delta\lambda$ with a mid-wavelength λ_0, the differential quotient $Q_\lambda(\lambda_0)$ is given by:

$$Q_\lambda(\lambda_0) = dQ(\delta\lambda)/d\lambda$$

In practice, instruments measure **difference quotients** for an interval $\Delta\lambda$ and the value represents an approach to the differential quotient at the mid-wavelength of the interval:

$$Q_\lambda(\Delta\lambda) = \Delta Q(\Delta\lambda)/\Delta\lambda \approx Q_\lambda(\lambda_0)$$

If energy or a derivative of energy, *e.g.* energy rate, is plotted as such against wavelength or another quantity indicating spectral position, the result is a cumulative function.

7.3.14   For the dimensionless derivatives of energy, such as internal transmittance and absorbance, and the coefficients derived from them, such as molar area absorbance (§9.86.3), there are two types of integration (for instance over an absorption band) for a quantity q:

1) $\int_{0\lambda(0)}^{1\lambda(1)} (q/\lambda)\mathrm{d}\lambda$ with dim q and unit $[q]$

2) $\int_{0\lambda(0)}^{1\lambda(1)} q\,\mathrm{d}\lambda$ with dim $q\,\lambda$ and unit $[q\,\lambda]$

The term *integrated absorbance*, for instance, is used for integrals over wavelength and has either dimension 1 or dimension L. In atomic absorption spectrometry, the same term *integrated absorbance* is used for an integral over time and has dimension T.[8]

7.4   Centrifugation

7.4.1   Centrifuges are widely used in clinical laboratories for separation of components. In biochemical analysis of body fluids, they are routinely used to separate blood cells from plasma and to separate formed elements from urine. They are used to measure the volume fraction of erythrocytes in blood and to separate bound from free components in protein binding and in immunochemical procedures. They are also used to separate lipoproteins in reference procedures for their measurement, to separate fractions of cells, and to distinguish fragments of deoxyribonucleates. The advice here on calculations and the information about kinds-of-quantity (Sections 8 and 9) are based on a recommendation of IUPAC-CQUCC and IFCC-CQU.[9] The IFCC Committee on Analytical Systems has also published guidelines for the selection and safe use of centrifuges, with related nomenclature, kinds-of-quantity and units.[10]

7.4.2   **Centrifugal acceleration**, a_{rot} (§9.34.1.1), may be calculated from centrifugal radius, r (§9.1.2.2) and **rotational frequency**, f_{rot} (§9.18.1.3):

$$a_{\mathrm{rot}} = 4\,\pi^2\,r\,f_{\mathrm{rot}}^2$$

It is commonly expressed in terms of **standard acceleration**, g_{n} (§9.34U), $g_{\mathrm{n}} = 9.806\,65\ \mathrm{m\ s^{-2}}$; but typical values can be expressed in $\mathrm{km\ s^{-2}}$. Rotational frequency is commonly expressed in 'revolutions per minute' (equivalent to $\mathrm{min^{-1}}$) or 'revolutions per second' (equivalent to $\mathrm{s^{-1}}$). Abbreviations such as rev/min, r.p.m., rev/s and rps are not recommended. It is recommended that scales of new equipment should be marked in hertz (Hz) (§9.18U).
Example of calculation

$$r = 170\ \mathrm{mm} = 0.17\ \mathrm{m}$$

$$f_{\mathrm{rot}} = 3000\ \mathrm{min^{-1}} = 50\ \mathrm{s^{-1}} = 50\ \mathrm{Hz}$$

$$a_{\mathrm{rot}} = 4\,\pi^2 \times 0.17\ \mathrm{m} \times (50\ \mathrm{Hz})^2$$

$$\approx 39.48 \times 0.17 \times 2500\ \mathrm{m\ s^{-2}}$$

$$\approx 16\,779\ \mathrm{m\ s^{-2}}$$

$$\approx 1711\ g_{\mathrm{n}}$$

7.4.3   The **kinetic energy**, E_{k} (§9.42.1.1), of a rotating body may be calculated by summation (or integration) of all partial masses, m_i, of the body at radii r_i from the axis of rotation:

$$E_{\mathrm{k}} = 2\,\pi^2\,f_{\mathrm{rot}}^2 \sum_{1}^{n} (m_i r_i^2)$$

$$= 2\,\pi^2\,f_{\mathrm{rot}}^2\,I$$

where I is moment of inertia (§9.11.1).

For a uniform disk of mass m

$$E_k = \pi^2 \ m \ r^2 \ f_{rot}^2$$

For a uniform ring of outer radius r and inner radius r_i

$$E_k = \pi^2 \ m \ r^2 \ f_{rot}^2 \ [1 - (r_i/r)]$$

7.5 Electrophoresis

7.5.1 Electrophoretic procedures are widely used in clinical laboratories to separate many types of component, including proteins, lipoproteins, isoenzymes and deoxyribonucleates. They are covered by a recommendation of IUPAC-CQUCC and IFCC-CQU.[11] A typical calculation is given (§7.5.2) and relevant kinds-of-quantity are listed in Sections 8 and 9.

7.5.2 **Electrophoretic mobility** (also termed velocity of migration; rate of migration) of an ionic (*i.e.* electrically charged) component, μ_B (§9.55.1), under specified conditions, including pH, substance concentration of ionic charge entity (or ionic strength), travel duration, and a given support medium, may be calculated as follows:

$$\mu_B = d \ l/(t \times U)$$

where d is distance travelled (§9.1.1.6), l is distance between anodic and cathodic sides of the strip (§9.1.1), t is travel duration (§9.17.1) and U is electric potential difference (§9.59.2).

Example of calculation

$$d_{(albumin)} = 25 \text{ mm} = 0.025 \text{ m}$$

$$l = 100 \text{ mm} = 0.1 \text{ m}$$

$$t = 1 \text{ h} = 3600 \text{ s}$$

$$U = 250 \text{ V}$$

$$\mu_{(albumin)} = (0.025 \text{ m} \times 0.1 \text{ m})/(3600 \text{ s} \times 250 \text{ V})$$

$$\approx 2.8 \times 10^{-9} \text{ m}^2 \text{ s}^{-1} \text{ V}^{-1} \text{ (preferred mode of expression)}$$

$$\approx 2800 \text{ mm}^2 \text{ s}^{-1} \text{ V}^{-1}$$

7.6 Enzymology

7.6.1 An extensive kind-of-quantity, catalytic amount, and an intensive kind-of-quantity, catalytic concentration, were defined in proposals of 1964 and published as recommendations in 1967. In view of the enormous debate these recommendations raised, a position paper was published.[12]

7.6.2 **Catalytic activity** (or *catalytic amount*) of an enzymic component E, z_E (§9.94.3), is defined from the increase in substance rate of reaction product (also termed *rate of conversion*) in the presence of the catalyst (§9.94.2.1). Substance rate has the SI unit mole per second; catalytic activity may be expressed in katals (kat) or moles per second. The concept is generalized and more tightly defined than the older concept of *enzyme activity* (or *enzym(at)ic activity*). Catalytic activity is proportional to the usually unknown number or amount-of-substance of **active centres** (or *active sites*) of the enzyme in the biological system under study.

7.6.3 Catalytic activities cannot be compared unless the conditions of measurement are the same. For instance, a catalytic activity of a given enzyme measured at a Celsius temperature of 30 °C cannot be used to calculate the activity at 37 °C, even if all other conditions are the same. The relationship may only be predicted by an empirical factor or by a regression equation, great care being needed in such empirical comparison because of any unknown accelerators, inhibitors or inter-actions present at different concentrations in different types of material and in different calibrators. In other words, different measurement procedures imply different kinds-of-quantity.

7.6.4 Results are commonly expressed by the intensive kind-of-quantity, **catalytic-activity concentration** (or **catalytic concentration**, for short), b_E (§9.97.2), and sometimes by **catalytic-activity content** (or **catalytic content**), z_E/m_1 (§9.98.2).

7.6.5 Whereas the **catalysed substance rate** of reactant B, $\xi_{B,cat}$ (§9.94.2.1), pertains to a reaction in a specified measuring system, **catalytic activity**, z_E, may be considered as a property of the enzyme in the original biological system.

7.6.6 The †enzyme unit, U (sometimes termed 'international unit'), was defined as 'the amount of enzyme that catalyses the transformation of one micromole of substrate per minute under standard conditions':[13]

$$U = \mu mol\,min^{-1}$$

The symbol U, when used in this sense, must not be confused with the symbol IU (meaning International Unit), which is the symbol used by the WHO (and recognized by CGPM in the SI Brochure) in expressing biological activity of certain substances that cannot yet be defined in terms of units of the SI. IUPAC-CQUCC and IFCC-CQU[14] recommended that the U be progressively replaced by submultiples of a unit (the katal, §7.6.7) coherent with SI (§9.94U).

7.6.7 The unit **katal**, symbol kat, was accepted by IUPAC, IFCC and IUB[15,16] as a special term for the derived unit mole per second, to be used only for catalytic activity. A request to the International Bureau of Weights and Measures for official recognition as an SI unit was finally granted by CGPM in 1999.[17] The unit katal, like other units apart from arbitrary units, *i.e.* procedure defined unit (p.d.u.), is by definition independent of the measurement procedure in contrast to the quantities expressed in that unit (ref. 18 and §7.6.3).

7.6.8 In the calculation of the catalytic-activity concentration of an enzyme according to the IFCC primary reference measurement procedure for lactate dehydrogenase,[19] it is assumed that the substance rate of reactant in the absence of enzyme is zero:

$$b_E = [V_1/(\varepsilon_B\,l\,V_2)] \times (\Delta_0^t A/\Delta t)_{cat}$$

where V_1 is volume of mixture of reactants including blood serum (in cuvette), V_2 is volume of blood plasma taken, ε_B is molar area decadic absorbance for an absorbing component derived from the reaction (§9.84.1), l is path length of a radiation through the mixture of reactants, A is decadic absorbance (§8.9.4.3), $\Delta_0^t A$ is change in decadic absorbance from time zero to time t, and Δt is the corresponding time difference.
Example of calculation for the catalytic-activity concentration of lactate dehydrogenase (b_{LDH}) according to IFCC[19]
When

$$V_1 = 2.3\ mL$$

$$\varepsilon_{NADH(340)} = 630\ m^2\,mol^{-1}$$

$$l = 0.01\ m$$

$$V_2 = 0.1\ mL, then$$

$$b_{LDH} = b_E = [2.3/(630 \times 0.01 \times 0.1)] \times (\Delta_0^t A/\Delta t)_{cat}$$

where b_{LDH} is expressed in µkat L^{-1}, volumes in mL, path length in metres, and molar area decadic absorbance of NADH $= 630$ m^2 mol^{-1}. A constant factor relates the change in decadic absorbance and the catalytic-activity concentration when the above conditions apply.

References

1. WHO, 1977. *The SI for the health professions*, prepared at the request of the Thirtieth World Health Assembly, WHO, Geneva, 1997.

2. P. J. Mohr, B. N. Taylor and D. B. Newell, Codata recommended values of the fundamental physical constants., *Rev. Mod. Phys.*, 2012, **84**, 1527–1605. It replaces the previously recommended 2006 CODATA set. http://physics.nist.gov/cuu/Constants/codata.pdf (acc. 2015-12).

3. O. Siggaard-Andersen, R. A. Durst and A. H. J. Maas, IUPAC-CQUCC & IFCC-EPpH 1984. Physicochemical quantities and units in clinical chemistry with special emphasis on activities and activity coefficients, *Pure Appl. Chem.*, 1984, **56**(5), 567–594. Also *J. Clin. Chem. Clin. Biochem.*, 1987, **25**(6), 369–391; *Ann. Biol. Clin.*, 1987, **45**, 89–109; *Labmedica*, 1986, **3**(6), 33–37.

4. N. Sheppard, H. A. Willis and J. C. Rigg, Names, Symbols, and Definitions of Quantities and Units in Optical Spectroscopy, *Pure Appl. Chem.*, 1985, **57**(1), 105–120. Also *J. Clin. Chem. Clin. Biochem.*, 1987, **25**(5), 327–33; *Labmedica*, 1987, **4**(2), 29–39; *Quím. Clín.*, 1988, **7**, 183–192.

5. R. Herrmann and C. Onkelinx, IUPAC-CQUCC; IFCC-EPQU, 1986. Quantities and units in clinical chemistry. Nebulizer and flame properties in flame emission and absorption spectrometry, *Pure Appl. Chem.*, 1986, **58**(12), 1737–1742; *Clin. Chim. Acta*, 1985, 48, 277F–285F; *J. Clin. Chem. Clin. Biochem.*, 1985, **23**, 365–371; *Medilab*, 1985, (8-9), 17–24.

6. M. S. Epstein, T. D. Geary, G. Gower, W. Tausch, K. J. Mills and D. Polt, IFCC-EPI, 1982. Provisional guidelines for listing specifications of atomic absorption spectrometers. IFCC Document Stage 2, Draft 1, *Clin. Chim. Acta*, 1982, **122**, 117F–123F. Also *J. Clin. Chem. Clin. Biochem.*, 1982, **20**, 263–266. Es: Pautas provisorias para el listado de especificaciones de espectrometros de absorcion atomica. *Acta Bioquim. Clín. Latinoam.*, 1983, **17**(1), 129.; *Clin. Chem. Newsletter* 1983, **3**, 3–13.

7. G. Bechtler, R. Haeckel, H. Lappe and W. Tausch, IFCC-EPI, 1984, Guidelines (1984) for listing specifications of spectometers in clinical chemistry. Stage 2, Draft 2, *J. Clin. Chem. Clin. Biochem.*, 1984, **22**, 565–571. Supersedes version of *Clin. Chim. Acta*, 1980, 103, 249F–258F; *J. Clin. Chem. Clin. Biochem.*, 1980, **18**, 947–951.

8. T. A. M. Ure, L. R. P. Butler, B. V. L'vov, I. Rubeska and R. Sturgeon, IUPAC-CSOPA, 1992. Nomenclature, symbols, units and their usage in spectrochemical analysis – XII. Terms related to electrothermal atomization (IUPAC Recommendations 1992), *Pure Appl. Chem.*, 1992, **64**(2), 253–259.

9. M. Lauritzen, IUPAC-CQUCC; IFCC-CQU, Quantities and units for centrifugation in the clinical laboratory, *Clin. Chim. Acta*, 1992, **205**, S5–S15. Also *Ann. Biol. Clin.*, 1992, **50**, 45–49; *J. Autom. Chem.*, 1992, **14**(4), 93–96; *Pure Appl. Chem.*, 1994, **66**(4), 897–907; *Acta Bioquim. Clín. Latinoam.*, 1993, **27**, 267–273.

10. A. Uldall, P. T. Damgaard, O. Drachmann, F. Jørgensen, D. Kennedy, M. Lauritzen, E. Magnussen, J. C. Rigg and P. Voss, IFCC-CAS, Guidelines (1990) for selection of safe laboratory centrifuges and for their safe use with general purpose appendices concerning centrifuge nomenclature, quantities and units, and calculation of centrifugal acceleration, *Clin. Chim. Acta*, 1991, **202**, S23–S40. Also *J. Autom. Chem.*, 1991, **13**(5), 221–229; *Ann. Biol. Clin.*, 1991, **49**, 439–447. Es: Pautas de IFCC (1990) para la seleccion de centrifugas de laboratorio seguras y para su uso seguro. Apéndice A y B. *Acta Bioquim. Clín. Latinoam.*, 1991, **25**, 447–461.

11. G. Férard, IUPAC-CQUCC; IFCC-CQU, Quantities and units for electrophoresis in the clinical laboratory, *J. Int. Fed. Clin. Chem.*, 1992, **4**(3), 122–126. Also: *Clin. Chim. Acta* 1992, **205**, 517–523; *Ann. Biol. Clin. Chem.*, 1992, **50**, 51–54; *J. Autom. Chem.* 1992,

14, 1–4; *Acta Bioquim. Clin. Chem. Latinoam.*, 1992, **26**, 455–460; *Pure Appl. Chem.* 1994, **66**(4):891–896.

12. R. Dybkær, Problems of quantities and units in enzymology, *Enzyme*, 1975, **20**, 46–64.

13. IUB. *Report of the Commission on Enzymes*, 1961, Pergamon Press, Oxford.

14. R. Dybkaer, IUPAC-CQUCC & IFCC-CQU (1979). List of quantities in clinical chemistry; Recommendation 1978, *Pure Appl. Chem.*, 1979, **51**(12), 2481–2502. Also *Clin. Chim. Acta*, 1979, **96**, 185F-204F.

15. IUB Nomenclature Committee. Units of Enzyme Activity, Recommendations 1978. *Eur. J. Biochem.*, 1979, **97**, 319–320.

16. IUPAC-IUB Joint Commission on Biochemical Nomenclature (JCBN), and Nomenclature Commission of IUB (IUB-NC), Newsletter 1981, *Eur. J. Biochem.*, 1981, **114**, 1–4; *Arch. Biochem. Biophys.*, 1981, **206**, 458–462; *Hoppe-Seyler's Z. Physiol. Chem.*, 1981, **362**, I–IV; *J. Biol. Chem.*, 1981, **256**, 12–14.

17. *Le Système International d'Unités (SI), The International System of Units (SI), 8th edn. (Bureau International des Poids et Mesures, Sèvres, France, 2006); it is the official French text followed by an English translation; commonly called the BIPM SI Brochure.* Available on the Web site of the BIPM: www.bipm.org/en/si/ (acc. 2015-12).

18. *IUPAC Section on Clinical Chemistry, Commission on Quantities and Units in Clinical Chemistry, IFCC Committee on Standards, Expert Panel on Quantities and Units.* Approved Recommendation (1978). List of Quantities in Clinical Chemistry. *Clin. Chim. Acta*, 1979, **96**, 157F–183F.

19. G. Schumann, R. Bonora, F. Ceriotti, C. A. Ferrero, G. Férard, P. F. Franck, F. J. Gella, W. Hoelzel, P. J. Jørgensen, T. Kanno, A. Kessner, R. Klauke, N. Kristiansen, J. M. Lessinger, T. P. Linsinger, H. Misaki, M. Panteghini, J. Pauwels, H. G. Schimmel, A. Vialle, G. Weidemann and L. Siekmann, IFCC primary reference procedures for the measurement of catalytic activity concentrations of enzymes at 37 degrees C. Part 3. Reference procedure for the measurement of catalytic concentration of lactate dehydrogenase, *Clin. Chem. Lab. Med.*, 2002, **40**(6), 643–648.

SECTION 8

Kinds-of-quantity of Dimension One; SI Unit 1

The many kinds-of-quantity of dimension one have been classified into 10 sections. After the (non-SI) base kind-of-quantity **number of entities** (§8.1.1), §8.2–10 are subdivided into various sorts of fractions, ratios and relative kinds-of-quantity (§5.4). Logarithmic kinds-of-quantity (§8.9) and exponential kinds-of-quantity (§8.10) are placed at the end of the section. The conventions on use of subscripts are explained in §4.2.8 and §4.3.14.

Each section is divided into data elements identified by numbers in angle brackets, whose meanings are defined in §3.6 and §3.8, and on the reference page xxiii (Structure of Numbered Entries in Sections 8 and 9).

A logical classification of relative kinds-of-quantity is more contentious than of fractions and ratios, which can almost always be defined from extensive kinds-of-quantity.

Example

- ☐ A kind-of-quantity defined from substance concentration is classed with those defined from amount-of-substance:

$$(c_B/c_R)_1 = (n_B/V_1)/(n_R/V_1) = (n_B/n_R)_1$$

  The same step can be taken for relative kinds-of-quantity only if the extensive kind-of-quantity of the system for numerator and denominator are the same.

Example

- ☐ Relative mass concentration, $(m_B/V_1)/(m_B/V_0)$, cannot be simplified to $m_{B,1}/m_{B,0}$ unless V_1 equals V_0 and then

$$[(m_B/V_1)/(m_B/V_0)]_V = (m_{B,1}/m_{B,0})_V$$

  However, for convenience, arrangement of kinds-of-quantity is such that, for instance, relative volumic mass, relative mass concentration and relative molar mass are all listed under §8.4: Mass fraction, mass ratio and relative derivatives of mass.

The heading of each section is a systematic term, usually constructed according to principles described in §5.4. No preference is expressed between the systematic term in that heading and other current terms indicated as full synonyms in bold type. Usages that are known to be obsolete or obsolescent are marked by a dagger (†).

At the end of each group of kinds-of-quantity is a subsection 'U' on relevant units (*e.g.* §8.1 ends with §8.1U).

Compendium of Terminology and Nomenclature of Properties in Clinical Laboratory Sciences: Recommendations 2016
Edited by Georges Férard, René Dybkaer and Xavier Fuentes-Arderiu
© International Union of Pure and Applied Chemistry 2017
Published by the Royal Society of Chemistry, www.rsc.org

As indicated in §5.3.3, redundant units such as 'mmol/mol' and 'µmol/mol' are accepted when the use of the undeclared unit 1 for a substance fraction results in an unwieldy number (that is, a numerical result violating the principle that the preferred interval for the measured quantity value should be between 0.1 and 999) (see §8.7U for example).

Note that ISO/IEC 80000-1[1] accepts the unit per cent (US percent), symbol % for the unit 0.01 (see §8.2U, §8.3U, §8.4U and §8.7U), and per mille (‰) for 0.001, but the use of part(s) per thousand (ppt), part(s) per million (ppm), part(s) per hundred million (pphm) and part(s) per billion (ppb) is discouraged (see §5.3.2).

8.1 Base Kinds-of-quantity of Dimension One

According to VIM [ref. 2, concept 1.8, Note 4] numbers of entities are kinds-of-quantity of dimension one and in quantity calculus, the kind-of-quantity 'number of entities' is often considered to be a base kind-of-quantity, with the base unit one, symbol 1.

8.1.1 ⟨2⟩ **number of entities** of component B in system 1
⟨3⟩ **absolute frequency** (statistics); number of individuals; number of particles; number of molecules or other elementary entities in a system
⟨4⟩ num.entities
⟨5⟩ N_B
⟨8⟩ count
⟨9⟩ (1) The **entity** counted may be any type of particle or mixture of particles, formula units or visible bodies, in other words any chemically or physically definable type of entity (§4.3).
⟨9⟩ (2) In statistics, **entities** are termed **items** or **individuals**, which may be members of a 'population, lot, sample, class, *etc.*'[3]
⟨10⟩ U—Erythrocytes; num.(proc.) $= 2.7 \times 10^6$
⟨10⟩ Sem(ejaculate)—Spermatozoa; num.(proc.) $= 2.1 \times 10^6$

8.1.1.1 ⟨2⟩ **number of entities** in sample
⟨5⟩ n
⟨8⟩ sample size

8.1.1.2 ⟨2⟩ **stoichiometric number** of reactant or product B
⟨5⟩ ν_B
⟨6⟩ number occurring in the equation of a chemical reaction
⟨8⟩ stoichiometric coefficient
⟨9⟩ (1) The stoichiometric numbers of a chemical reaction are specified by the reaction equation. Their values are negative for reactants and positive for products. Values depend on how the reaction equation is written.
⟨9⟩ (2) A chemical reaction may be symbolized $0 = \sum\limits_{B}^{N} (\nu_B B)$

⟨9⟩ (3) Equilibrium of a chemical reaction is characterized by:

$$0 = \sum_{B}^{N} (\nu_B \mu_B) \quad \text{and} \quad 0 = \prod_{B}^{N} \lambda$$

⟨9⟩ (4) For the reaction $N_2 + 3H_2 = 2NH_3$

$$\nu(N_2) = -1/2 = -0.5$$

$$\nu(H_2) = -3/2 = -1.5$$

$$\nu(NH_3) = 2/2 = +1$$

8.1.1.3 $\langle 2\rangle$ **charge number** of ionic component B or of an ion B
$\langle 5\rangle$ z_B
$\langle 6\rangle$ electrical charge of an ion of the component, Q_B, divided by the elementary charge of a proton, e
$\langle 7\rangle$ $z_B = Q_B/e$

8.1U **Units of number of entities**
$\langle 12\rangle$ *Coherent SI unit* one (1)

Factor	Term for unit				Value in coherent SI unit
	Recommended	Other	Long scale	Short scale	
10^{18}			trillion	pentillion	$= 1 \times 10^{18}$
10^{15}			billiard	quadrillion	$= 1 \times 10^{15}$
10^{12}			billion	trillion	$= 1 \times 10^{12}$
10^{9}	milliard			billion	$= 1 \times 10^{9}$
10^{6}	million				$= 1 \times 10^{6}$
10^{3}	thousand				$= 1 \times 10^{3}$
10^{2}	hundred				$= 0.1 \times 10^{3}$
10^{1}	ten				$= 1 \times 0.01 \times 10^{3} = 1 \times 10$
1	(unit) **one**	unity			$= 1$

$\langle 19\rangle$ (1) The long and short scales are two different systems for terming large numbers. Most English-speaking countries use the short scale and most of the countries in continental Europe use the long scale.

$\langle 19\rangle$ (2) For the problems of expressing very large or very small quantities of dimension one see §5.3.3.

$\langle 19\rangle$ (3) The formulations such as 1×10^{18} and $1 \times 0.01 \times 10^{3}$ in the last column are given to show the relation to the coherent unit one (1).

8.2 Number Fraction and Number Ratio

8.2.1 $\langle 2\rangle$ **number fraction** of component B in group of entities B, C, ... , N in system 1
$\langle 3\rangle$ particle fraction, entity fraction
$\langle 4\rangle$ num.fr.
$\langle 5\rangle$ δ_B
$\langle 6\rangle$ (1) number of entities of a defined type constituting the component divided by the total number of defined entities in the system

$$\langle 7\rangle \ \delta_B = N_B \Bigg/ \sum_B^N N$$

$\langle 6\rangle$ (2) number of times that an outcome falls into a particular interval (or a particular class) divided by the total number of outcomes (statistical usage)
$\langle 8\rangle$ relative frequency; frequency; proportion; fraction; f
$\langle 9\rangle$ (1) The term *particle fraction* is too restrictive. The type of entity may or may not be particles.
$\langle 9\rangle$ (2) For chemical entities, the value of number fraction equals the value of substance fraction. In cytometry and in particle size distribution, only number fraction is applicable.
$\langle 10\rangle$ Ercs(B)—Reticulocytes; num.fr. $= 6 \times 10^{-3} = 6$ ‰ $= 0.6$ %
$\langle 10\rangle$ Sperm—Spermatozoa(motile with IgA); num.fr.(MAR; proc.) $= 0.15 = 15$ %

8.2.1.1 $\langle 2\rangle$ **number fraction of component B expected** in system 1
$\langle 3\rangle$ **probability** (in a predictive sense)
$\langle 5\rangle$ P, p

8.2.1.2 $\langle 2 \rangle$ **number fraction of molecules of component B dissociated**
$\langle 3 \rangle$ **degree of dissociation**; dissociation fraction
$\langle 5 \rangle$ α
$\langle 6 \rangle$ number of dissociated molecules of a component divided by the stoichiometric number of undissociated and dissociated molecules of the component
$\langle 10 \rangle$ P(B)—Calcium ionized/Calcium total; num.fr. $= 0.42 = 42$ %

8.2.1.3 $\langle 2 \rangle$ **number fraction of people infected** with pathogen B
$\langle 3 \rangle$ **prevalence of a given infection**; proportion of people infected
$\langle 5 \rangle$ $\delta_{B,inf}$
$\langle 10 \rangle$ Population X—VIH infected people; prevalence $= 0.041 = 4.1$ %
$\langle 12 \rangle$ Population X must be defined according to, for example, geographical, ethnic or age criteria, . . .

8.2.1.4 $\langle 2 \rangle$ **number fraction of people** with disease B
$\langle 3 \rangle$ **prevalence of disease B**
$\langle 5 \rangle$ $\delta_{B,dis}$
$\langle 6 \rangle$ number of people suffering from a given disease divided by the total number of people in a population
$\langle 10 \rangle$ Population X—Diabetic people; prevalence $= 0.124 = 12.4$ %
$\langle 12 \rangle$ Population X must be defined, for example, in geographical, ethnical or ageing terms.

8.2.1.5 $\langle 2 \rangle$ **number fraction of a specific type of cells within a class of cells** in the same system
$\langle 6 \rangle$ Number of cells of a particular type divided by the total number of cells of a given class within a system
$\langle 10 \rangle$ B—Eosinophils/Leukocytes; num.fr.(visual counting) $= 0.05 = 5$ %
$\langle 10 \rangle$ Seminal spermatozoa—Spermatozoa(vital); num.fr.(proc.) $= 0.48 = 48$ %
$\langle 10 \rangle$ Spermatozoa—Progressive motility; num.fr.(proc.) $= 0.41 = 41$ %
$\langle 9 \rangle$ The concept 'motility' is defined in sperm examination.

8.2.2 $\langle 2 \rangle$ **number ratio** of component B to component C in system 1
$\langle 4 \rangle$ num.ratio
$\langle 5 \rangle$ $N_{r,B/C}$
$\langle 6 \rangle$ number of entities of component B in system divided by the number of entities of component C in the same system
$\langle 10 \rangle$ Bone marrow—Myeloid cells/Erythroid cells; num.ratio(proc.) $= 0.40 = 40$ %

8.2U **Units of number fraction and number ratio**
$\langle 12 \rangle$ *Coherent SI unit* one (1)

Term for unit	Symbol of unit Recommended	Value in coherent SI unit
(unit) **one**	1	$= 1$
per cent	%	$= 1 \times 10 \times 10^{-3} = 0.01$
ten to power minus three	10^{-3}	$= 1 \times 10^{-3} = 0.001$
per mille	‰	$= 1 \times 10^{-3} = 0.001$
ten to power minus six	10^{-6}	$= 1 \times 10^{-6}$
ten to power minus nine	10^{-9}	$= 1 \times 10^{-9}$
ten to power minus twelve	10^{-12}	$= 1 \times 10^{-12}$

$\langle 19 \rangle$ (1) For the problems of expressing unwieldy numerical values of quantities of dimension one see §5.3.3.
$\langle 19 \rangle$ (2) ISO/IEC 80000-1[1] states that abbreviations such as ppt, ppm, pphm and ppb shall not be used.

8.3 Volume Fraction, Volume Ratio and Volume Rate Ratio

8.3.1   ⟨2⟩ **volume fraction** of component B in system 1

⟨4⟩ vol.fr.

⟨5⟩ φ_B, ϑ_B

⟨6⟩ (1) volume of the isolated component, V_B, divided by the volume of the system, V_1

⟨7⟩ $\varphi_B = V_B/V_1$

⟨6⟩ (2) volume of the isolated component, V_B, divided by the total volume of the separate components of the system, $\sum\limits_{A}^{N} V_i$

⟨7⟩ $\varphi_B = V_B \Big/ \left(\sum\limits_{A}^{N} V_i \right)$

⟨8⟩ concentration (ambiguous)

⟨9⟩ (1) Definition 1 applies in preparative chemistry but can also be derived from the substance fractions and the partial molar volumes of the components. Definition 2 is identical to Definition 1 only if mixing of the components causes no change in volume at a given temperature and pressure. The second definition can also be derived from the product of substance fraction and molar volume of each of the pure substances in the system.

⟨9⟩ (2) For components whose elementary entity can be defined in chemical terms, substance concentration (§9.88.1) is a more informative kind-of-quantity.

⟨10⟩ B—Erythrocytes; vol.fr. $= 0.43 = 43$ % or 0.43 L/L

8.3.2   ⟨2⟩ **volume ratio** of component B to component C in system 1

⟨4⟩ vol. ratio

⟨5⟩ $V_{r,B/C}$

8.3.2.1 ⟨2⟩ **volume ratio** of component B to solvent component A

⟨5⟩ $V_{r,B/A}$

⟨9⟩ A statement of dilution or dilution strength, for instance, as 1:2 or 1:3 is ambiguous and could mean either volume ratio or volume fraction. Volume fraction of each component is preferred in specifications of made-up solutions, together with the preparative procedures. See also ISO 78-2.[4]

⟨10⟩ P(B)—Lipids/Water; vol. ratio $= 0.047 = 4.7$ %

8.3.3   ⟨2⟩ **volume rate ratio** of component B to component A in system 1

⟨4⟩ vol. rate ratio

⟨10⟩ Kidn—Albumin clearance/Creatininium clearance; vol.rate ratio $= 0.01 = 1$ % or (10 μL/s)/(1 mL/s)

8.3U   **Units of volume fraction, volume ratio and volume rate ratio**

⟨12⟩ *Coherent SI units* one or cubic metre(s) per cubic metre ($m^3\ m^{-3} = 1$)

Term for unit	Symbol of unit		Value in coherent SI unit
	Recommended	Other	
(unit) **one**	1		$= 1$
cubic metre(s) per cubic metre	$m^3\ m^{-3}$		$= 1$
litre(s) per litre	$L\ L^{-1}, l\ l^{-1}$		$= 1$
millilitre(s) per millilitre	$mL\ mL^{-1}, ml\ ml^{-1}$		$= 1$
per cent	%		$= 1 \times 10 \times 10^{-3} = 0.01$
ten to power minus three	10^{-3}		$= 1 \times 10^{-3} = 0.001$
cubic centimetre(s) per cubic decimetre		$cm^3\ dm^{-3}$	$= 1 \times 10^{-3} = 0.001$
millilitre(s) per litre	$mL\ L^{-1}, ml\ l^{-1}$		$= 1 \times 10^{-3} = 0.001$

(Continued)

Term for unit	Symbol of unit		Value in coherent
	Recommended	Other	SI unit
per mille	‰		$= 1 \times 10^{-3} = 0.001$
ten to power minus six	10^{-6}		$= 1 \times 10^{-6}$
microlitre(s) per litre	$\mu L\ L^{-1}$, $\mu l\ l^{-1}$		$= 1 \times 10^{-6}$
ten to power minus nine	10^{-9}		$= 1 \times 10^{-9}$
cubic millimetre(s) per cubic metre	$mm^3\ m^{-3}$		$= 1 \times 10^{-9}$
nanolitre(s) per litre	$nL\ L^{-1}$, $nl\ l^{-1}$		$= 1 \times 10^{-9}$

⟨19⟩ ISO/IEC 80000-1[1] and CGPM[5] deprecate symbols such as % (*V/V*), ‰ (*V/V*), ppmV (part per million by volume); description of the kind-of-quantity belongs to the left side of the equation where the measurand is defined.

8.4 Mass Fraction, Mass Ratio and Relative Derivatives of Mass

8.4.1 ⟨2⟩ **mass fraction** of component B in system 1
⟨4⟩ mass fr.
⟨5⟩ w_B
⟨6⟩ (1) mass of a component, m_B, divided by the mass of the system, m_1
⟨7⟩ $w_B = m_B/m_1$
⟨6⟩ (2) mass of a component, m_B, divided by the total mass of a series of components, $\sum_A^N m_i$, constituting the system

⟨7⟩ $w_B = m_B \Big/ \left(\sum_A^N m_i \right)$

⟨8⟩ content
⟨9⟩ For components whose elementary entity can be defined, substance content (§9.91.1) is a more informative kind-of-quantity.
⟨10⟩ Faes(dry)—Lipid(total); mass fr. $= 0.157 = 15.7$ %
The result may also be expressed as 157 g kg^{-1}.
⟨10⟩ Prot(U)—Alpha-2-globulin; mass fr.(proc.) $= 0.05 = 5$ % $= 50$ mg g^{-1}

8.4.2 ⟨2⟩ **mass ratio** of component B to component R in system 1
⟨5⟩ $m_{r,B/R}$
⟨8⟩ † weight ratio; w
⟨9⟩ A mass concentration ratio of two components, $\rho_{r,B/R}$, equals their mass ratio as long as they are both in the same volume of system and not in different volumes of subsystems (§5.4.4(2))
⟨10⟩ P—Albumin/Globulin; mass ratio(electrophoresis) $= 1.40 = 140$ %

8.4.3 ⟨2⟩ **relative volumic mass** of system 1 to system 0
⟨3⟩ relative mass density
⟨4⟩ rel.volumic mass
⟨5⟩ d_1, $\rho_{r,1}$
⟨6⟩ volumic mass of a system under specified conditions, ρ_1, divided by the volumic mass of a reference system, ρ_0, under the same conditions or under different specified conditions
⟨7⟩ $d_1 = \rho_1/\rho_0$
⟨8⟩ relative density; †specific gravity; †specific weight
⟨9⟩ (1) The specifications of such a quantity should include the terms of both systems.

⟨9⟩ (2) Conditions on which relative volumic mass of a system depends are temperature, which must always be stated, and, for gases, also pressure.

⟨9⟩ (3) The term †"specific gravity" was used where the reference system was water, usually at 4 °C, and the system under consideration at 20 °C, with the symbol d_4^{20}. Until 1964, values of specific gravity equalled numerical values of mass density (volumic mass) in kilograms per litre (1901). Continued use of old data on specific gravity in tables of data is still causing systematic errors because the litre was redefined in 1964 (§9.5U, Table and Note 1).

⟨10⟩ B—Plasma; rel.volumic mass(20 °C/water, 4 °C) = 1.026 = 102.6 %

8.4.4   ⟨2⟩ **relative mass concentration** of component B in system 1 to system 0

⟨4⟩ rel.mass c.

⟨5⟩ $\rho_{r,B}$

⟨6⟩ mass concentration of a component in a given system divided by the mass concentration of the same component in another system, usually considered as a reference system.

⟨10⟩ Air—Water vapour; relative mass concentration(temperature t; actual/maximum).

⟨9⟩ Synonym: **relative humidity**; h

⟨10⟩ CsF—Albumin; rel.mass c.(CsF/P) = 6.2×10^{-3} = 6.2 ‰

8.4.5   ⟨2⟩ **relative molar mass** of component B

⟨4⟩ rel.molar mass

⟨5⟩ $M_{r,B}$

⟨6⟩ molar mass of a component divided by one-twelfth of the molar mass of carbon-12

⟨7⟩ $M_{r,B} = M_B/[M(^{12}C)/12]$
$= (m_B/N_B)/[m_{at}(^{12}C)/12]$,   where m_{at} is the atomic mass.

⟨8⟩ atomic weight ($A_{r,B}$); molecular weight; relative atomic (or molecular) mass; AW, MW

⟨9⟩ (1) Relative molar mass of a component B equals the mass of that component divided by the number of entities and by one-twelfth of the mass of an atom of carbon-12 (also termed the unified atomic mass constant or the Dalton) (Table 5.3).

⟨9⟩ (2) For atoms, the term 'atomic weight' is accepted.

⟨9⟩ (3) See also §7.1.5.

⟨10⟩ Human—Albumin; rel.molar mass = 66 437

8.4U   **Units of mass fraction, mass ratio and relative derivatives of mass**

⟨12⟩ *Coherent SI units* one or kilogram(s) per kilogram (kg kg^{-1} = 1) or (kg L^{-1})/(kg L^{-1}) = 1

Term for unit	Symbol of unit		Value in coherent SI unit
	Recommended	Other	
(unit) **one**	1		= 1
kilogram(s) per kilogram	kg kg^{-1} or		= 1
	(kg L^{-1})/(kg L^{-1})		= 1
gram(s) per gram		g g^{-1}	= 1
per cent	%		= $1 \times 10 \times 10^{-3}$ = 0.01
ten to power minus three	10^{-3}		= 1×10^{-3} = 0.001
gram(s) per kilogram	g kg^{-1}		= 1×10^{-3} = 0.001
milligram(s) per gram		mg g^{-1}	= 1×10^{-3} = 0.001
per mille	‰		= 1×10^{-3} = 0.001
ten to power minus six	10^{-6}		= 1×10^{-6}
milligram(s) per kilogram	mg kg^{-1}		= 1×10^{-6}
microgram(s) per gram		μg g^{-1}	= 1×10^{-6}

(Continued)

Term for unit	Symbol of unit		Value in coherent SI unit
	Recommended	Other	
ten to power minus nine	10^{-9}		$= 1 \times 10^{-9}$
microgram(s) per kilogram	$\mu g\ kg^{-1}$		$= 1 \times 10^{-9}$
nanogram(s) per gram		$ng\ g^{-1}$	$= 1 \times 10^{-9}$
ten to power minus twelve	10^{-12}		$= 1 \times 10^{-12}$
nanogram(s) per kilogram	$ng\ kg^{-1}$		$= 1 \times 10^{-12}$
picogram(s) per gram		$pg\ g^{-1}$	$= 1 \times 10^{-12}$

$\langle 19 \rangle$ (1) For problems of expressing unwieldy numerical values of quantities of dimension one see §5.3.3.

$\langle 19 \rangle$ (2) ISO/IEC 80000-1 [1] and CGPM [5] deprecate symbols such as % (m/m), ‰ (m/m).

8.5 Relative Time and Relative Derivatives of Time

8.5.1 $\langle 2 \rangle$ **relative time**
$\langle 5 \rangle$ t_r
$\langle 6 \rangle$ time divided by another time (§5.4.4.3)
$\langle 10 \rangle$ P—Coagulation, time factor-induced; rel.time(actual/norm; INR; IRP67/40; proc.) $= 0.80 = 80$ %

8.5.2 $\langle 2 \rangle$ **relative lineic time for radiation** λ of non-absorbing medium 1
$\langle 3 \rangle$ **refractive index** of non-absorbing medium 1 to radiation λ
$\langle 5 \rangle$ n
$\langle 6 \rangle$ speed of a radiation in vacuum, c_0, divided by the speed of the radiation in a medium, c_1
$\langle 7 \rangle$ $n = c_0/c_1$
$\langle 9 \rangle$ (1) For equal distance travelled in a uniform medium, refractive index equals a relative lineic time:

$$n = (\mathrm{d}l/\mathrm{d}t)_0/(\mathrm{d}l/\mathrm{d}t)_1$$

$$= (\mathrm{d}t/\mathrm{d}l)_1/(\mathrm{d}t/\mathrm{d}l)_0$$

$$= (t_1/t_0)_l$$

$\langle 9 \rangle$ (2) In most regions of the spectrum, the difference in refractive index between air and a vacuum is negligible and hence refractive index is usually calculated as the sine of the angle of incidence of a beam in air, α_0, divided by the sine of the angle of refraction within the medium, α_1

$$n = \sin \alpha_0/\sin \alpha_1$$

$\langle 9 \rangle$ (3) The definition is valid as long as the medium does not absorb radiation. For an absorbing medium, the relative lineic time is a complex quantity, $\hat{n}$ with a real part n.

8.5U **Units of relative time and relative derivatives of time**
$\langle 12 \rangle$ *Coherent SI units* one or second(s) per second ($s\ s^{-1} = 1$)

8.6 Energy Fraction

8.6.1 $\langle 2 \rangle$ **energy fraction** of component B in system 1
$\langle 6 \rangle$ energy of a component, $\tau_{i,B}$, divided by the total energy, τ_i, of the system

$\langle 7 \rangle$ $Q_B \Big/ \sum\limits_{A}^{N} Q_i$

⟨9⟩ This kind-of-quantity is often used in physiology and nutrition.

⟨10⟩ Person—Energy intake(Fat)/Energy consumed(total); energy fr.(1 d) $= 0.30 = 30$ % or 0.30 J J^{-1}

8.6.1.1
⟨2⟩ **energy fraction of radiation** λ **absorbed**

⟨3⟩ **internal absorptance; internal absorption factor**

⟨5⟩ $\alpha_i(\lambda)$, $\alpha(\lambda)$

⟨8⟩ absorptance; absorption factor; radiant absorptance; absorption

⟨9⟩ (1) The energy fraction of a radiation lost from a parallel beam during passage through a medium cannot be equated with absorptance, because it may also be partly due to dissipation (*i.e.* scattering and reflection). Only if other causes of loss are excluded can the definition of absorptance be simplified and transposed:

$$P_{\lambda,\text{lost}} = P_{\lambda,0} - P_{\lambda,\text{tr}}$$

If $P_{\lambda,\text{lost}} = P_{\lambda,\text{abs}}$

$$\alpha_i(\lambda) = P_{\lambda,0} - P_{\lambda,\text{tr}}/P_{\lambda,0}$$

where $P_{\lambda,0}$ is the energy rate for the radiation incident on the system so that

$$\alpha_i(\lambda) = 1 - \tau_i(\lambda) \quad \text{where } \tau_i \text{ is the energy of radiation.}$$

⟨9⟩ (2) **Internal**, symbolized by subscript i, assumes the exclusion of boundary effects, for instance at the surface of the cuvette, so that the decrease in energy in the direction of the beam is entirely due to processes (absorption and scatter) within the medium (§7.3.4–5).

⟨9⟩ (3) Such a quantity varies with position in the spectrum. The radiation may be defined by wavelength, wavenumber or frequency, symbolized $\alpha_i(\lambda)$, $\alpha_i(\tilde{\nu})$ or $\alpha_i(\nu)$, respectively. For a sufficiently narrow spectral interval, the three functions correspond exactly (§7.3.12–13).

8.6U Units of energy fraction
⟨12⟩ *Coherent SI units* one or joule(s) per joule (J J$^{-1} = 1$)

8.7 Substance Fraction, Substance Ratio, Relative Amount-of-substance and Relative Substance Concentration

8.7.1
⟨2⟩ **substance fraction** of component B in system 1

⟨3⟩ mole fraction; amount-of-substance fraction

⟨4⟩ subst.fr.

⟨5⟩ x_B, y_B

⟨6⟩ amount-of-substance of a component, n_B, divided by the total amount-of-substance of all specified entities, $\sum\limits_{A}^{N} n_i$, in the system

⟨7⟩ $x_B = n_B \bigg/ \sum\limits_{A}^{N} n_i$

$= n_B/(n_A + n_B +, \dots , + n_N)$

⟨8⟩ number fraction; amount fraction

⟨9⟩ (1) In contrast to substance concentration, substance fraction is independent of temperature and pressure. It is usually more informative than mass fraction, volume fraction, substance concentration or molality, and is therefore useful where all the components can be specified.

⟨9⟩ (2) For principles necessary to specify components and entities see §4.3.

⟨9⟩ (3) For a condensed phase (*i.e.* a liquid or a solid), the symbol x is used and the solvent component A is usually excluded from the summation. The system is then the sum of a set of components considered and the condensed phase is taken as a supersystem.

⟨10⟩   P(B)—Calcium   ion(II)/Calcium(total);   subst.fr.(proc.) $= 0.45 = 45$ % $=$ 450 mmol mol^{-1}

⟨10⟩ Hb(Fe; B)—Carbon monoxide haemoglobin(Fe); subst.fr.(proc.) $= 0.09 = 9$ % $= 90$ mol mol^{-1}

8.7.2   ⟨2⟩ **substance ratio of component B to component C** in system 1
⟨3⟩ mole ratio; amount-of-substance ratio
⟨4⟩ subst.ratio
⟨5⟩ $r_{B/C}$
⟨6⟩ amount-of-substance of a component, n_B, in the system divided by the amount-of-substance of another component, n_C, in the same system
⟨7⟩ $r_{B/C} = n_B/n_C$
⟨10⟩ U—Adenosyl-L-methionine/Creatininium; subst.ratio $= 0.10 = 10$ % $= 100$ mol mol^{-1}
⟨10⟩ Sweat—Potassium ion/Sodium ion; subst.ratio $= 0.31 = 31$ % $= 31$ mol mol^{-1}

8.7.3   ⟨2⟩ **relative amount-of-substance** of component B in system 1 to that in a reference system
⟨4⟩ rel.am.s.
⟨7⟩ $n_{B,1}/n_{B,0}$
⟨6⟩ amount-of-substance of component in a system 1 divided by a reference amount-of-substance of the same component in another reference system
⟨10⟩ U—Xylose; rel.am.s.(U1 d/uptake; proc.) $= 0.26 = 26$ %

8.7.4   ⟨2⟩ **relative amount-of-substance** of component B in system 1 to that in a saturated system
⟨4⟩ rel.am.s.
⟨5⟩ s_B, S_B
⟨6⟩ amount-of-substance of a component in a system divided by the amount-of-substance in a saturated system
⟨7⟩ $s_B = [n_B/n_{B,sat}]_{T,p,n(A),n(C),\ldots,\,n(N)}$
⟨8⟩ saturation fraction (ambiguous)
⟨10⟩ P(a)—Oxygen; rel.am.s(actual/saturated; proc.) $= 0.90 = 90$ %

8.7.5   ⟨2⟩ **relative substance concentration** of component B in system 1 to that in a normal system
⟨4⟩ rel.subst.c.
⟨6⟩ substance concentration of a component B in a system divided by the substance concentration in a normal system
⟨10⟩ P—Coagulation factor XI; rel.subst.c.(imm.; actual/norm; proc.) $= 0.70 = 70$ %

8.7U   **Units of substance fraction, substance ratio, relative amount-of-substance and relative substance concentration**
⟨12⟩ *Coherent SI units* one or mole(s) per mole (mol mol$^{-1} = 1$)

Term for unit	Symbol of unit Recommended	Value in coherent SI unit
(unit) **one**	1	$= 1$
mole(s) per mole	mol mol^{-1}	$= 1$
per cent	%	$= 10 \times 10^{-3} = 0.01 \times 1$
ten to power minus three	10^{-3}	$= 10^{-3} = 0.001$
millimole(s) per mole	mmol mol^{-1}	$= 10^{-3} = 0.001$

(*Continued*)

Term for unit	Symbol of unit Recommended	Value in coherent SI unit
per mille	‰	$= 10^{-3} = 0.001$
ten to power minus six	10^{-6}	$= 10^{-6}$
micromole(s) per mole	μmol mol^{-1}	$= 10^{-6}$

⟨19⟩ For problems of expressing unwieldy numeric values of measurable quantities of dimension one see §5.3.3.

8.8 Catalytic-activity Fraction

⟨2⟩ **catalytic-activity fraction** of isoenzyme E_i in enzyme E of system 1
⟨4⟩ cat.fr.
⟨5⟩ $z_{f,i}$
⟨6⟩ catalytic activity of an isoenzyme, z_i, divided by the catalytic activity of all isoenzymes of the enzyme, z_E, in the system
⟨7⟩ $z_{f,i} = z_i/z_E$

$$= z_i \Big/ \sum_1^N z$$

⟨10⟩ CK(P)—Creatine kinase MB; cat.fr.(37 °C; proc.) $= 0.11 = 11$ %
⟨10⟩ ChE(P)—Choline esterase, cinchocaine inhibited; cat.fr.(37 °C; proc.) $= 0.30 = 30$ %

8.8U **Units of catalytic-activity fraction**
⟨12⟩ *Coherent SI units* one or katal(s) per katal (kat kat$^{-1} = 1$)

8.9 Logarithmic Kinds-of-quantity

⟨9⟩ (1) Where a phenomenon, a process, is described in terms of decadic (or Briggsian) logarithms, the modifier **decadic** may be added. The term **attenuance** was coined specifically for the decadic variant in order to avoid the lengthy specifications.
⟨9⟩ (2) Binary logarithms are also sometimes used, often unknowingly (*e.g.* half-life, half-length), and may be described with the modifier **binary**.
⟨9⟩ (3) In some branches of science, napierian, decadic, and binary logarithmic kinds-of-quantity have been considered as variants with different '*units of dimension one*'.
⟨9⟩ (4) The kind-of-quantity $\Delta Q/Q$ should be termed **fractional difference**, or **fractional change** (or *relative change)*, except for a small difference, δQ, which is much less than Q. The kind-of-quantity $\delta Q/Q$ approaches $\delta \ln Q$.

8.9.1 ⟨2⟩ **napierian absorbance** of radiation λ in uniform medium 1
⟨5⟩ $B(\lambda)$
⟨6⟩ negative napierian logarithm of (one minus internal absorptance)
⟨7⟩ $B(\lambda) = -\ln [1 - \alpha_i(\lambda)]$

8.9.2 ⟨2⟩ **attenuance** of radiation λ in uniform medium 1
⟨5⟩ $D_i(\lambda), D(\lambda)$
⟨6⟩ negative decadic logarithm of internal transmittance of a parallel beam of a radiation λ through that medium
⟨7⟩ $D_i(\lambda) = -\lg \tau_i(\lambda) = -\lg [1 - \alpha_i(\lambda)]$
⟨8⟩ extinction; optical density; optical transmission density; transmission density

⟨9⟩ (1) Attenuance applies without specification of the cause of the decrease in the radiation transmitted but assumes the internal property without surface effects, *i.e.* it includes losses by absorption or scattering. By contrast, optical density and the symbol D includes also surface effects, for instance in application to blackening of images in photographs.

⟨9⟩ (2) For a system with scatter, some of the scattered radiation reaches the detector. How much does so depends on the geometry of the optical system used for collection of the radiation.

8.9.3 ⟨2⟩ **decadic absorbance** of radiation λ in uniform medium 1
⟨5⟩ $A(\lambda)$
⟨6⟩ negative decadic logarithm of (one minus radiant absorptance)
⟨7⟩ $A(\lambda) = -\lg [1 - \alpha_i(\lambda)]$
⟨8⟩ absorbance; $A_i(\lambda)$
⟨9⟩ (1) In the absence of scatter with $\alpha_i(\lambda) = [1 - \tau_i(\lambda)]$,

$$A(\lambda) = -\lg \tau_i(\lambda) = \lg [1/\tau_i(\lambda)] = \lg (P_{\lambda,0}/P_{\lambda,\mathrm{tr}})$$

⟨9⟩ (2) In contrast to attenuance, the term **absorbance** applies only to the decrease in transmitted radiation due to absorption. Both attenuance and absorbance apply only to the internal property, excluding surface effects.

⟨9⟩ (3) In absorption spectrometry of solutions, the reference differential quotient of energy rate to wavelength, $P_{\lambda,0}$, is not that of incident radiation but that of the radiation transmitted by pure solvent or some other reference system.

⟨10⟩ For an example of calculation of catalytic concentration see §7.6.8.

8.9.4 ⟨2⟩ **pH**
⟨3⟩ **hydrogen ion exponent**
⟨5⟩ pH
⟨6⟩ negative decadic logarithm of the relative molal activity of hydrogen ion, $a_b(\mathrm{H}^+)$
⟨7⟩ $\mathrm{pH} = -\lg a_b(\mathrm{H}^+)\tilde{b}$
        $= -\lg [\gamma(\mathrm{H}^+) \, b(\mathrm{H}^+)/\tilde{b}_\ominus]$
        $= -\{[\mu(\mathrm{H}^+) - \mu_\ominus]/(R \, T \ln 10)\}$

The molal standard reference system is chosen such that

$$\gamma_\ominus \, b_\ominus = \tilde{b}_\ominus = 1 \text{ mol kg}^{-1}$$

⟨9⟩ (1) The concept of pH was first introduced by Sørensen (1909) as

$$\mathrm{pH} = -\lg c(\mathrm{H}^+)/c_\ominus \quad \text{where } c_\ominus = 1 \text{ mol kg}^{-1}.$$

The difference between definitions based on substance concentration and on molality in dilute aqueous solutions can be ignored for most purposes; the definition is valid when $\mathrm{pH} = [2, 12]$. The relative difference depends on the volumic mass of water and amounts to about 0.001 at ambient temperatures.

⟨9⟩ (2) The quantity is linearly related to the **chemical potential** of hydrogen ion, $\mu(\mathrm{H}^+)$.

⟨9⟩ (3) The definition in terms of the effect of hydrogen ion can only be notional, because the effects of a single species of ion cannot be separated from that of its counterion. In other words, the electrochemical potential cannot be distinguished into a chemical and an electrical component. Distinction must necessarily be based on a non-thermodynamic convention.[6]

⟨9⟩ (4) The present convention (Bates–Guggenheim convention) is based on the assumption that the relative molal activity of the chloride ion in dilute aqueous solutions ($I_b < 0.10$ mol kg^{-1}) can be estimated by means of the Debye–Hückel equation:

$$-\lg \gamma_\mathrm{B} = z_\mathrm{B}^2 \, A \, I^{0.5}/(1 + \mathring{a} \, B \, I_b^{0.5})$$

where I_b is ionic strength, z_B is charge number of the ion, $\mathring{a}$ is a coefficient of dimension L for the ion, and A and B are temperature-dependent constants.

According to the convention of Bates and Guggenheim (1960), $\mathring{a}\,B$ is taken as 1.5 mol$^{0.5}$ kg$^{-0.5}$ at all temperatures and for all compositions of the solutions.

⟨9⟩ (5) The present convention is based on the use of an electrode responsive to hydrogen ions, a reference electrode and a bridge solution of concentrated KCl of molality ≥ 3.5 mol kg^{-1}.

⟨9⟩ (6) It has often been suggested that substance concentration of hydrogen ion should be reported instead of pH. However, because of the international agreement on the pH scale and the reference method for pH measurement[7] recommends maintenance of use of the quantity pH, also in clinical chemistry (§7.2.10).

⟨9⟩ (7) The pH concept should not be generalized to other species of ion (*e.g.* Na$^+$, K$^+$, Ca^{2+}) measured with ion-selective electrodes, or to any other components. Ion-selective electrodes should, however, be calibrated in a manner analogous to pH electrodes, *i.e.* on the basis of relative molal activity (a_b). The results may be reported as substance concentration (§9.88.1) of the ion.

⟨10⟩ P(aB)—Hydrogen ion; pH(37 °C) = 7.40
Do not use pH = 7.40 units.

8.9.5   ⟨2⟩ **iso-electric point** of component B in solution 1
⟨5⟩ pH(I)
⟨6⟩ pH at which the net electrical charge of an entity of the component is zero
⟨8⟩ pI
⟨9⟩ The iso-ionic point of a component is the pH at which the net electrical charge of the component in pure water is zero.
⟨10⟩ Person—Albumin; iso-electric point = 4.9

8.9.6   ⟨2⟩ **negative decadic logarithm of apparent equilibrium coefficient** of reaction B in system 1
⟨5⟩ $-\lg K'_B$
⟨8⟩ pK'_B
⟨9⟩ In this instance, the requirement that the kind-of-quantity must be of dimension one is fulfilled if the equilibrium coefficient is defined in terms of substance fractions. Otherwise the quantity can only be calculated as $-\lg (K'_B/K_\ominus)$.
⟨10⟩ B(Person)—Reaction $CO_2 + H_2O \rightleftharpoons H^+ + HCO_3^-$; negative decadic apparent substance-fractional equilibrium coefficient = 6.105

8.9.7   ⟨2⟩ **negative decadic logarithmic number concentration of entities** in system 1
⟨9⟩ Biological logarithmic concentrations such as viral loads which are often reported as decadic log(number concentration)
⟨10⟩ P—Hepatitis C virus RNA; negative decadic logarithmic num.c.(proc.) = 2.3

8.9U   **Units of logarithmic kinds-of-quantity**
⟨12⟩ *Coherent SI unit* one (1)
⟨12⟩ *Coherent scale* napierian logarithmic scale

| Term for unit | Symbol of unit | | Value in coherent SI unit |
	Recommended	Other	
(unit) **one**	1		= 1
absorbance unit		A.U.	= 1
'pH unit'		—	= 1
bel		B	= 1
neper		Np	= 1
decibel		dB	= 0.1

⟨19⟩ The symbol A.U. is discouraged.

8.10 Exponentially Defined Kinds-of-quantity

8.10.1 $\langle 2 \rangle$ **chemical activity** of component B in system 1
        $\langle 3 \rangle$ **absolute chemical activity**
        $\langle 5 \rangle$ λ_B
        $\langle 7 \rangle$ $\lambda_B = \exp\left[\mu_B/(R \times T)\right]$
        $\langle 8 \rangle$ activity; absolute activity
        $\langle 9 \rangle$ Only relative chemical activity can be measured (§7.2.7 and §8.10.2).

8.10.2 $\langle 2 \rangle$ **relative chemical activity** of component B in system 1
        $\langle 5 \rangle$ a_B
        $\langle 7 \rangle$ $a_B = \lambda_B/\lambda_{B,\ominus}$
              $= \exp\left[(\mu_B - \mu_{B,\ominus})/(R \times T)\right]$
        $\langle 8 \rangle$ relative activity
        $\langle 9 \rangle$ See §7.2.7.

8.10.2.1 $\langle 2 \rangle$ **relative chemical activity** of component A in system 1
        $\langle 5 \rangle$ a_A
        $\langle 7 \rangle$ $a_A = \lambda_A/\lambda_{A,\ominus}$ where the standard state is for pure solvent or major component
                                   $(x_A = 1)$.
             $= \lambda_A/\lambda_{A*}$ where the star indicates 'pure'.
             $= f_A\, x_A$
        $\langle 9 \rangle$ Component A may be solvent or the major component of a mixture.

8.10.2.2 $\langle 2 \rangle$ **relative molal activity** of component B in system 1
        $\langle 5 \rangle$ $a_{b,B}$, $a_{m,B}$
        $\langle 7 \rangle$ $a_{b,B} = \lambda_B/\lambda_{B,\ominus}$ where the standard state is $b_{B,\ominus} = 1$ mol kg^{-1}
                $= \tilde{b}_B/\tilde{b}_{B,\ominus}$
        $\langle 8 \rangle$ molal activity
        $\langle 9 \rangle$ (1) This is the relative chemical activity commonly used in clinical chemistry for an ionic component in solution with an ion-selective electrode.
        $\langle 9 \rangle$ (2) For a strong electrolyte, where $B \rightarrow \nu_C\,C + \nu_D\,D$, the standard reference system is defined by:

$$b_{\pm,B,\ominus} = 1 \text{ mol kg}^{-1} \; (\S 8.10.3.2, \langle 9 \rangle \, (2))$$

It follows that

$$a_{b,B} = a_{b,C}{}^{\nu(C)}\, a_{b,D}{}^{\nu(D)}$$
$$= a_{b,\pm,B}{}^{[\nu(C)+\nu(D)]}$$

8.10.2.3 $\langle 2 \rangle$ **relative substance-concentrational activity** of component B in system 1
        $\langle 5 \rangle$ $a_{c,B}$
        $\langle 7 \rangle$ $a_{c,B} = \lambda_B/\lambda_{B,\ominus}$ where the standard state is $c_{B,\ominus} = 1$ mol kg^{-1}
               $= \tilde{c}_B/\tilde{c}_{B,\ominus}$
        $\langle 8 \rangle$ concentrational relative activity; activity

8.10.2.4 $\langle 2 \rangle$ **relative substance-fractional activity** of component B in system 1
        $\langle 5 \rangle$ $a_{x,B}$
        $\langle 7 \rangle$ $a_{x,B} = \lambda_B/\lambda_{B,\ominus}$ where the standard state is $f_{x,B,\ominus}\, x_{B,\ominus} = 1$
             $= f_{x,B}\, x_B$
        $\langle 8 \rangle$ rational activity; relative rational activity; relative activity; activity
        $\langle 9 \rangle$ $a_{x,B}/\tilde{b}_B = M_A$

8.10.3 $\langle 2 \rangle$ **activity factor** of component B in system 1
        $\langle 5 \rangle$ γ_B
        $\langle 8 \rangle$ activity coefficient

8.10.3.1 ⟨3⟩ **activity factor** of solvent component A in solution 1

⟨5⟩ f_A

⟨7⟩ $f_A = (\lambda_A/x_A)/\lim_{x(A)\to 1} (\lambda_A/x_A)$

$= (\lambda_A/\lambda_{A*})/x_A$ where the star indicates 'pure'.

⟨8⟩ activity coefficient

⟨9⟩ The limiting value for pure solvent is 1.

8.10.3.2 ⟨2⟩ **molal activity factor** of component B for solvent A in solution 1

⟨5⟩ $\gamma_{b,B},\ \gamma_{m,B}$

⟨7⟩ $a_{b,B} = \gamma_{b,B}\ b_B/b_\ominus$

$= (\lambda_B/b_B)/\lim_{x(A)\to 1} (\lambda_B/b_B)$

⟨8⟩ molal activity coefficient, activity coefficient; γ_B

⟨9⟩ (1) $\lim_{x(A)\to 1} \gamma_{b,B} = \gamma_{b,B,\infty} = 1$

⟨9⟩ (2) A strong electrolyte B dissociating into cation C and anion D is described in terms of mean ionic quantities indicated by subscript $\pm$. For instance:

$$\gamma_{b,\pm,B} = \left(\gamma_{b,C}^{\nu(C)}\ \gamma_{b,D}^{\nu(D)}\right)^{1/[\nu(C)+\nu(D)]}$$

⟨9⟩ (3) Corresponding variants exist for chemical activity, molality and molal relative activity.

8.10.3.3 ⟨2⟩ **substance-concentrational activity factor** of component B in system 1

⟨5⟩ $\gamma_B,\ \gamma_{c,B}$

⟨7⟩ $y_B = (\lambda_B/c_B)/\lim_{x(A)\to 1} (\lambda_B/c_B)$

⟨8⟩ substance-concentrational activity coefficient

⟨9⟩ (1) $\lim y_B = 1$

⟨9⟩ (2) $\gamma_B/y_B = \rho_A/\rho_{A*}$ where the star indicates 'pure'.

8.10.3.4 ⟨2⟩ **substance-fractional activity factor** of solute B in system 1

⟨5⟩ $f_{x,B},\ \gamma_{x,B}$

⟨7⟩ $f_{x,B} = (\lambda_B/x_B)/\lim_{x(A)\to 1} (\lambda_B/x_B)$

⟨8⟩ rational activity coefficient

⟨9⟩ The limiting value at infinite dilution may be designated $f_{x,B,\infty}$ (§7.2.7).

8.10.4 ⟨2⟩ **osmotic factor** of solution 1

⟨5⟩ φ

⟨8⟩ osmotic coefficient

8.10.4.1 ⟨2⟩ **molal osmotic factor** of solute component B in solution 1

⟨5⟩ $\varphi_b,\ \varphi_m$

⟨6⟩ osmolality of a solute component divided by the molality of all the solute components B, ... ,N in the system

⟨7⟩ $\varphi_{b,B} = \widehat{b}_B \Big/ \sum_B^N b_B$

$= (\mu_{A*} - \mu_A) \Big/ \left(RTM_A \sum_B^N b\right)$ where the star indicates 'pure'

$= -\left(M_A \sum_B^N b\right)^{-1} \ln a_A$

⟨8⟩ molal osmotic coefficient

⟨9⟩ (1) $\sum_B^N b_B = (1 - x_A)/(x_A M_A)$

⟨9⟩ (2) $\lim_{x(A)\to 1} \varphi_{b,B} = \varphi_{b,B,\infty} = 1$

8.10.4.2 $\langle 2 \rangle$ **substance-concentrational osmotic factor** of solution 1

$\langle 5 \rangle$ φ_c

$\langle 7 \rangle$ $\varphi_c = \widehat{c} \bigg/ \sum\limits_{B}^{N} c_B$

$\qquad = \varphi_B \, \rho_{A*}/\rho_A$ where the star indicates 'pure'.

$\langle 8 \rangle$ concentrational osmotic coefficient

$\langle 9 \rangle$ (1) $\sum\limits_{B}^{N} c_B = c_A(1 - x_A)/x_A$

$\langle 9 \rangle$ (2) $\lim_{x(A) \to 1} \varphi_c = \varphi_{c,\infty} = 1$

8.10.4.3 $\langle 2 \rangle$ **substance-fractional osmotic factor**

$\langle 5 \rangle$ φ_x

$\langle 7 \rangle$ $\varphi_x = x \bigg/ \sum\limits_{B} x$ where x is the 'osmotic substance fraction' defined as $x = -\ln a_A$

$\langle 8 \rangle$ substance-fractional osmotic coefficient

8.10.5 $\langle 2 \rangle$ **partial order of reaction** for reactant B

$\langle 5 \rangle$ $n_B, n(B)$

$\langle 7 \rangle$ $v = k \, \Pi c_B{}^{n(B)}$

References

1. ISO/IEC 80000-1:2009/ Cor 1:2011. *Quantities and units — Part 1: General.* Supersedes ISO 31-0: 1992 with the original title *Quantities and units — Part 1: General principles.*

2. BIPM; IEC; IFCC; ILAC; ISO; IUPAC; IUPAP; OIML, 2012. *International vocabulary of metrology – Basic and general concepts and associated terms VIM*, 3rd edn. JCGM 200: 2012. This 3rd edition is also published as ISO Guide 99 by ISO (ISO/IEC Guide 99-12: 2007). Replaces the 2nd edition, 1993. Available on the Web site of the BIPM: http://www.bipm.org. (acc. 2015-12).

3. ISO 3534-2:2006. *Statistics — Vocabulary and symbols — Part 2. Statistical quality control.*

4. ISO 78-2:1983. *Standard for chemical analysis.* Superseded by ISO 78-2:1999. *Chemistry — Layouts for standards — Methods of chemical analysis.*

5. *Le Système International d'Unités (SI), The International System of Units (SI)*, 8th edn. (Bureau International des Poids et Mesures, Sèvres, France, 2006); it is the official French text followed by an English translation; commonly called the BIPM SI Brochure. Available on the Web site of the BIPM: www.bipm.org/en/si/ (acc. 2015-12).

6. R. P. Buck, S. Rondinini, A. K. Covington, F. G. K. Baucke, C. M. A. Brett, M. F. Camões, M. J. T. Milton, T. Mussini, R. Naumann, K. W. Pratt, P. Spitzer and G. S. Wilson, Measurement of pH. Definition, standards, and Procedures (IUPAC Recommendations 2002), *Pure Appl. Chem.*, 2002, **74**(11), 2169–2200.

7. O. Siggaard-Andersen, R. A. Durst and A. H. J. Maas, IUPAC-CQUCC and IFCC-EPpH, Physicochemical quantities and units in clinical chemistry with special emphasis on activities and activity coefficients, *Pure Appl. Chem.*, 1984, **56**(5), 567–594. Also *J. Clin. Chem. Clin. Biochem.*, 1987, **25**(6), 369–391; *Ann. Biol. Clin.*, 1987, **45**, 89–109; *Labmedica*, 1986, **3**(6), 33–37.

Kinds-of-quantity of Dimension Other than One

The many kinds-of-quantity of dimension other than one have been classified into 109 sections. The order of entries is explained in §3.5, based on the dimensions of SI base units apart from electrical and photometric kinds-of-quantity, which are arranged in order of derivation from the dimension of electrical charge (§9.50) and of amount-of-light (§9.79). Each section is subdivided into subsections related to detailed applications in clinical laboratory sciences. Properties outside the seven dimensions of SI are placed in Section 10. Each subsection is divided into data elements identified by numbers in angle brackets, whose meanings are defined in §3.6 and §3.9, and on the reference page xxiii (Structure of Numbered Entries in Sections 8 and 9). The conventions about use of subscripts are explained in §5.4.12. Abbreviations used in examples under ⟨10⟩ are those listed in Table 6.1 (for systems) and in Table 6.2 (for kinds-of-property).

9.1   Dimension L

9.1.1   ⟨2⟩ **length**
           ⟨3⟩ **distance** between two points 1 and 2
           ⟨5⟩ l, L
           ⟨10⟩ Faes—Parasite(spec.); length(average) = ? mm
           ⟨10⟩ B—Sedimentation reaction; length(proc.) = 12 mm

9.1.1.1   ⟨2⟩ **elongation**
           ⟨3⟩ **change in length**, length elongation
           ⟨5⟩ Δl
           ⟨10⟩ Pt(10 d postmenstrual)—Cervix mucus spinnbarkeit; length elongation = 2 mm

9.1.1.2   ⟨2⟩ **length of path travelled** by entity B
           ⟨3⟩ **path-length; path length**
           ⟨5⟩ s
           ⟨8⟩ path; L
           ⟨9⟩ Path travelled is not necessarily a straight line.

9.1.1.3   ⟨2⟩ **path length** of radiation λ in medium 1
           ⟨3⟩ **length traversed**
           ⟨5⟩ l
           ⟨8⟩ b

Compendium of Terminology and Nomenclature of Properties in Clinical Laboratory Sciences:
Recommendations 2016
Edited by Georges Férard, René Dybkaer and Xavier Fuentes-Arderiu
© International Union of Pure and Applied Chemistry 2017
Published by the Royal Society of Chemistry, www.rsc.org

9.1.1.4   ⟨2⟩ **linear coordinates**
          ⟨3⟩ **Cartesian space coordinates**; Cartesian coordinates
          ⟨5⟩ (x,y,z), x,y,z

9.1.1.5   ⟨2⟩ **linear displacement** of entity B in system 1
          ⟨5⟩ ξ
          ⟨6⟩ instantaneous distance of an entity from its position at rest
          ⟨7⟩ $\xi = (\delta x, \delta y, \delta z)$
          ⟨8⟩ displacement; x

9.1.1.6   ⟨2⟩ **distance** travelled or migrated (as a scalar quantity, *i.e.* a quantity having
          magnitude but no direction)
          ⟨5⟩ d

9.1.1.7   ⟨2⟩ **breadth**
          ⟨5⟩ b, y
          ⟨8⟩ width; w

9.1.1.8   ⟨2⟩ **height**
          ⟨5⟩ h, z
          ⟨10⟩ Pt—Body; height $= 1.76$ m

9.1.1.9   ⟨2⟩ **depth**
          ⟨3⟩ **thickness**
          ⟨5⟩ d, $-z$, δ
          ⟨10⟩ Person—Triceps skinfold; thickness(proc.) $= 12$ mm

9.1.2   ⟨2⟩ **radius of circle**
          ⟨5⟩ r, R

9.1.2.1   ⟨2⟩ **radius of curvature of arc**
          ⟨5⟩ ρ

9.1.2.2   ⟨2⟩ **centrifugal radius**
          ⟨5⟩ r
          ⟨6⟩ radius at which a component is spinning when the centrifuge is at equilibrium
          ⟨9⟩ For a component sedimented from a dilute suspension, it can be equated with
          the radius of rotation at the bottom of the centrifuge tube, as long as the volume
          fraction of the component in the suspension is small.

9.1.2.3   ⟨2⟩ **diameter of circle**
          ⟨5⟩ $2\,r$
          ⟨8⟩ d, D
          ⟨10⟩ B—Erythrocytes; average diameter $= 7.4$ μm
          ⟨10⟩ Skin(spec.)—Tomato induced papule; diameter(proc.) $= 6$ mm

9.1.3   ⟨2⟩ **entitic length** of entities of component B in system 1
          ⟨7⟩ l_1/N_B
          ⟨10⟩ Faes—*Enterobius vermicularis*(pinworm, female); entitic length $= 10$ mm

9.1.3.1   ⟨2⟩ **wavelength**
          ⟨5⟩ λ
          ⟨6⟩ distance in the direction of propagation of a regular wave, l, divided by the
          number of cycles of the wave in that distance, $N_\sim$
          ⟨7⟩ $\lambda = l/N_\sim$

⟨9⟩ (1) It can also be considered as the distance in the direction of propagation of a periodic wave between two points where the phase is the same at a given time.

⟨9⟩ (2) By convention, spectroscopic properties are expressed in terms of wavelength in vacuum, λ_0, not the wavelength within the medium being studied. The wavelength within the medium, λ_1, depends on its refractive index, n_1 (§8.5.2):

$$\lambda_1/\lambda_0 = n_0/n_1$$

By definition, the refractive index of a vacuum is 1. The refractive index of air at room temperature and pressure is very close to 1, so wavelength measured in air is valid for most practical purposes, except in spectral regions of strong atmospheric absorption.

⟨9⟩ (3) The micrometre is an appropriate unit for the infrared region and the nanometre for the visible and ultraviolet region (Table 7.3).

9.1.4 ⟨2⟩ **areic volume** of component B for interface 1
⟨5⟩ V_B/A_1
⟨6⟩ volume of a component divided by the area of a specified interface or surface

9.1U **Units of dimension L**
⟨12⟩ *SI base unit* metre(s) (m)
⟨16⟩ The metre is the length of the path travelled by light in vacuum during a time interval of 1/299 792 458 of a second (17th CGPM 1983, Resolution 1). The definition of the base units is currently being discussed and may be fixed on natural constants (see §5.1.9).

	Symbol of unit		
Term for unit	Recommended	Other	Value in coherent SI unit
kilometre(s)	km		$= 1000$ m
metre(s)	m		$= 1$ m
decimetre(s)		dm	$= 0.1$ m
centimetre(s)		cm	$= 0.01$ m
millimetre(s)	mm		$= 10^{-3}$ m
micrometre(s)	μm		$= 10^{-6}$ m
micron(s)		μ, u, mu	$= 10^{-6}$ m
nanometre(s)	nm		$= 10^{-9}$ m
millimicron(s)		mμ	$= 10^{-9}$ m
ångström		Å	$= 0.1 \times 10^{-9}$ m
picometre(s)	pm		$= 10^{-12}$ m
X-unit(s)		XU, X	$\approx 100.207\,72 \times 10^{-15}$ m
femtometre(s)	fm		$= 10^{-15}$ m

⟨19⟩ (1) The spelling metre is recommended by CGPM and by ISO. In the United States, the spelling meter is commonly used.

⟨19⟩ (2) The ångström has no official sanction from CIPM or CGPM, but is widely used by X-ray crystallographers and structural chemists.[1]

9.2 Dimension L^{-1}

9.2.1 ⟨2⟩ **lineic number** of component B in direction x,y,z
⟨3⟩ repetency
⟨5⟩ σ
⟨7⟩ $\sigma = \mathrm{d}(x,y,z)/\mathrm{d}N$

9.2.1.1   ⟨2⟩ **lineic number of waves**
⟨3⟩ **wavenumber**; wavenumber in a medium
⟨5⟩ σ
⟨6⟩ (1) number of cycles of a regular wave, N, in a given distance, l, in the direction of propagation, divided by that distance
⟨7⟩ $\sigma = N/l$
⟨6⟩ (2) reciprocal of wavelength, λ
⟨7⟩ $\sigma = 1/\lambda$
⟨9⟩ Wavenumber in vacuum is commonly used in defining a radiation. Wavenumber in the medium is rarely used in optical spectroscopy.

9.2.1.2   ⟨2⟩ **lineic number of waves in vacuum**
⟨3⟩ **wavenumber in vacuum**
⟨5⟩ $\tilde{\nu}$
⟨7⟩ $\tilde{\nu} = \nu/c_0$
        $= 1/n_1\,\lambda_1$
        $= 1/\lambda_0$ where ν is frequency, c_0 is speed of electromagnetic radiation in vacuum, n_1 is refractive index of a medium, λ_1 is wavelength in that medium, and λ_0 is wavelength in vacuum.
⟨9⟩ (1) Wavenumber in vacuum is commonly used in specifying a radiation. For instance in spectrometry, wavenumber is that in vacuum or in air, not that in the medium under investigation.
⟨9⟩ (2) Care is needed with the tilde in the symbol, because ν represents frequency and $\bar{\nu}$ represents average frequency.

9.2.2   ⟨2⟩ **attenuation coefficient**
⟨5⟩ κ
⟨7⟩ $-\mathrm{d}\ln Q/\mathrm{d}l$, μ_l
⟨9⟩ For general considerations about logarithmic kinds-of-quantity, see §8.9.

9.2.2.1   ⟨2⟩ **linear napierian attenuation coefficient to radiation** λ of system 1
⟨5⟩ $\mu(\lambda)$, $\mu_l(\lambda)$
⟨6⟩ (1) relative decrease in the differential coefficient of energy rate of a collimated beam of electromagnetic radiation to wavelength during traversal of an infinitesimal layer of a medium, divided by the length traversed
⟨7⟩ (1) $\mu(\lambda) = (\mathrm{d}P_\lambda/\mathrm{d}l)/P_\lambda$
⟨7⟩ (2) $\mu(\lambda) = -(\ln \tau_\mathrm{i})/l$ where τ_i is internal transmittance.
        $= B/l$
where τ_i is internal transmittance, l is path length of the beam through a uniform medium and B is napierian attenuation coefficient.
⟨8⟩ linear attenuation coefficient; linear extinction coefficient; attenuation coefficient; extinction coefficient

9.2.2.2   ⟨2⟩ **lineic attenuance to radiation** λ of system 1
⟨3⟩ linear decadic attenuation coefficient; linear decadic extinction coefficient
⟨5⟩ $m(\lambda)$
⟨6⟩ attenuance, D, divided by path length, l, of a parallel beam through a medium of uniform properties
⟨7⟩ $m(\lambda) = D(\lambda)/l$
        $= -[\lg \tau_\mathrm{i}(\lambda)]/l$ where τ_i is internal transmittance
⟨8⟩ linear attenuation coefficient; attenuation coefficient; extinction coefficient

9.2.2.3 ⟨2⟩ **lineic napierian absorbance** to radiation λ of system 1
⟨3⟩ linear napierian absorption coefficient; **linear napierian absorptivity**
⟨5⟩ $\alpha_l(\lambda)$
⟨6⟩ (1) that part of the linear attenuation coefficient that is due to absorption
⟨6⟩ (2) napierian absorbance, B, divided by path length, l, of a parallel beam through a medium of uniform properties
⟨7⟩ $\alpha_l(\lambda) = B(\lambda)/l$
$\qquad = -[\ln \tau_i(\lambda)]/l$
⟨8⟩ linear absorption coefficient; napierian absorption coefficient; $\alpha(\lambda)$; $a(\lambda)$
⟨9⟩ The symbol α is used also for absorptance. IUPAC has reserved the symbol a for the decadic equivalent.

9.2.2.4 ⟨2⟩ **lineic decadic absorbance** to radiation λ of system 1
⟨3⟩ **linear decadic absorption coefficient** of radiation due to absorption
⟨5⟩ a_l, a, K
⟨6⟩ (1) that part of the internal linear (decadic) attenuation coefficient which is due to absorption
⟨6⟩ (2) decadic absorbance, A, divided by the path length, l, of a parallel beam within a uniform medium
⟨7⟩ $a = A/l$
⟨8⟩ linear absorption coefficient; decadic absorption coefficient; absorption co-efficient; lineic absorbance; absorptivity
⟨9⟩ (1) If the cause of the loss of power is undefined, the kind-of-quantity is lineic attenuance.
⟨9⟩ (2) The alternative symbol K has the disadvantage that it is used also for luminous efficacy.
⟨9⟩ (3) Lineic decadic absorbances due to absorption by a series of components B, C, ... , N may be symbolized a_B, a_C, ... , a_N (§4.3.14) and, according to theory, should be additive as long as the components do not interact.

9.2.3 ⟨2⟩ **pH gradient** of system 1 at position (x,y,z)
⟨5⟩ **grad** pH, ∇pH
⟨6⟩ differential change in local pH with position, (x,y,z), divided by change in position
⟨7⟩ ∇pH $= dpH/\mathrm{d}(x,y,z)$
⟨9⟩ It is a vector kind-of-quantity.

9.2U **Units of dimension L^{-1}**
⟨12⟩ *Coherent SI units* (one) per metre, reciprocal metre(s), or metre(s) to the power minus one (m^{-1})

Term for unit	Symbol of unit		Value in coherent SI unit
	Recommended	Other	
per millimetre	mm^{-1}		$= 1000\ m^{-1}$
per centimetre		cm^{-1}	$= 100\ m^{-1}$
per decimetre		dm^{-1}	$= 10\ m^{-1}$
per metre	m^{-1}		$= 1\ m^{-1}$
neper(s) per metre		$Np\ m^{-1}$	$= 1\ m^{-1}$
dioptre(s)		dpt, d	$= 1\ m^{-1}$

⟨19⟩ The customary unit for wavenumber in optical spectroscopy is cm^{-1}.

9.3 Dimension L^2

9.3.1 ⟨2⟩ **area of plane, surface or interface**
⟨5⟩ A, S

$\langle 6 \rangle$ integral product of length, x, and breadth, y

$\langle 7 \rangle$ $A = \int^2 \mathrm{d}x\, \mathrm{d}y$

9.3.1.1 $\langle 2 \rangle$ **entitic area** of component B in system 1

$\langle 5 \rangle$ A_B/N_B

$\langle 6 \rangle$ area of a component divided by the number of specified entities of the component

$\langle 10 \rangle$ B—Erythrocyte surface; entitic area $= 140\ \mu m^2$

9.3U   **Units of dimension L^2**

$\langle 12 \rangle$ *Coherent SI unit* square metre(s) (m^2)

Term for unit	Symbol of unit Recommended	Other	Value in coherent SI unit
square kilometre(s)	km^2		$= 10^6\ m^2$
hectare(s)		ha	$= 10 \times 10^3\ m^2$
are(s)		a	$= 100\ m^2$
square metre(s)	m^2		$= 1\ m^2$
centiare(s)		ca	$= 1\ m^2$
square centimetre(s)		cm^2	$= 0.1 \times 10^{-3}\ m^2$
square millimetre(s)	mm^2		$= 10^{-6}\ m^2$
litre(s) per kilometre		$L\ km^{-1}, l\ km^{-1}$	$= 10^{-6}\ m^2$
square micrometre(s)	μm^2		$= 10^{-12}\ m^2$
square micron(s)		μ^2	$= 10^{-12}\ m^2$

9.4   Dimension L^{-2}

9.4.1 $\langle 2 \rangle$ **areic number** of component B for surface or interface 1

$\langle 5 \rangle$ $N_{A,B}$

$\langle 6 \rangle$ (1) (particle physics). Number of particles incident on a small sphere, N_B, at a given point in space, divided by the cross-sectional area of that sphere, A_1

$\langle 6 \rangle$ (2) (generalized). Number of entities related to a surface or crossing a surface, N_B, divided by the area of that surface or interface, A_1

$\langle 7 \rangle$ $N_{A,B} = N_B/A_1$

$\langle 8 \rangle$ particle fluence; density; surface coverage (used in surface chemistry); Φ

$\langle 10 \rangle$ Country—People; areic num. $= 110\ km^{-2}$

$\langle 9 \rangle$ Synonym: population density

$\langle 10 \rangle$ Skin—Petechiae(pressure induced); areic num.(proc.) $= 200\ m^{-2}$

9.4U   **Units of dimension L^{-2}**

$\langle 12 \rangle$ *Coherent SI units* (one) per square metre or reciprocal square metre(s) or metre(s) to the power minus two (m^{-2})

Term for unit	Symbol of unit Recommended	Other	Value in coherent SI unit
per square millimetre	mm^{-2}		$= 10^6\ m^{-2}$
kilometre(s) per litre		$km\ L^{-1}, km\ l^{-1}$	$= 10^6\ m^{-2}$
per square metre	m^{-2}		$= 1\ m^{-2}$
per square kilometre	km^{-2}		$= 10^{-6}\ m^{-2}$

9.5 Dimension L^3

9.5.1   ⟨2⟩ **volume**
      ⟨4⟩ vol.
      ⟨5⟩ V
      ⟨6⟩ integral product of length, x, breadth, y and height, z
      ⟨7⟩ (1) $V = \int^3 (dx\, dy\, dz)$
      ⟨7⟩ (2) For a cube, $V = l^3$
      ⟨8⟩ v
      ⟨9⟩ As measures of the 'size' of a system, amount-of-substance (§9.83.1), or mass (§9.8.1) are sometimes more informative (§7.1.2 and §7.1.5).
      ⟨10⟩ Pt—Blood; vol.(proc.) = 5.5 L
      ⟨10⟩ Pt—Semen ejaculate; vol.(proc.) = 4.1 mL
      ⟨10⟩ Pt—Stomach fluid; vol.(proc.) = 160 mL

9.5.2   ⟨2⟩ **entitic volume** of component B in system 1
      ⟨4⟩ entitic vol.
      ⟨5⟩ V_B/N_B
      ⟨6⟩ volume of a component divided by the number of specified entities of the component
      ⟨10⟩ B—Erythrocytes; entitic vol. = 90 fL

9.5U   **Units of dimension L^3**
      ⟨12⟩ *Coherent SI unit* cubic metre(s) (m³)

Term for unit	Symbol of unit Recommended	Other	Value in coherent SI unit
cubic metre(s)	m³		= 1 m³
decalitre(s)		daL, dal	= 0.010 m³
litre(s) (1901–1964)		l	≈ 1.000 028 × 10⁻³ m³
cubic decimetre(s)		dm³	= 0.001 m³
litre(s)	L, l		= 0.001 m³
decilitre(s)		dL, dl	= 0.1 × 10⁻³ m³
centilitre(s)		cL, cl	= 0.01 × 10⁻³ m³
cubic centimetre(s)		cm³, cc	= 10⁻⁶ m³
millilitre(s)	mL, ml		= 10⁻⁶ m³
cubic millimetre(s)	mm³		= 10⁻⁹ m³
microlitre(s)	μL, μl		= 10⁻⁹ m³
lambda(s)		λ	= 10⁻⁹ m³
nanolitre(s)	nL, nl		= 10⁻¹² m³
picolitre(s)	pL, pl		= 10⁻¹⁵ m³
cubic micrometre(s)	μm³		= 10⁻¹⁸ m³
femtolitre(s)	fL, fl		= 10⁻¹⁸ m³

⟨19⟩ (1) The litre was originally defined as now, in terms of the cubic decimetre. In 1901, it was redefined as the volume of 1 kilogram of water at maximum volumic mass (Celsius temperature about 4 °C) and standard pressure. This definition was never fully implemented and was withdrawn in 1964. However, some published tables of data based on volume by that definition still contain systematic errors through use of the wrong factor, for instance tables of 'density' (volumic mass) or of viscosity (Resolution 13 of the 11th CGPM 1960; BIPM 2006).[2]

⟨19⟩ (2) By resolution of the Commission on Clinical Chemistry of IUPAC and of IFCC,[3] the litre was retained for use in the clinical laboratory in expression of volumes and of substance concentrations or mass concentrations, in preference to the cubic metre. In expression of concentrations, submultiples of the litre should not be used (§5.10.2). The preference for the litre in clinical sciences was declared in 1972 by a recommendation of ICSH *et al.* (1973, Section 1),[1] and reaffirmed in the SI brochure.[2] The advantages and disadvantages noted then were as follows.

- The symbol l is easier to write than m^3 or than dm^3.
- The symbol l is easily confused with the numeral 1. For this reason, the 16th General Conference on Weights and Measures (1979) resolved to permit the alternative symbol L as an exception to general practice of allowing only one symbol.[2] In this Compendium, the symbol L is preferred.
- The usual decimal symbols may be prefixed to the symbol for litre, giving a convenient series in steps of 10^3, whereas prefixes to the cubic metre give steps of 10^9.

9.6   Dimension L^{-3}

9.6.1 $\langle 2 \rangle$ **volumic number** of entities B in system 1
$\langle 4 \rangle$ volumic num.
$\langle 5 \rangle$ n_B

$\langle 6 \rangle$ total number of entities in a system, $\sum\limits_{A}^{N} N$, divided by the volume of the system, V_1

$\langle 7 \rangle$ $n_B = \left(\sum\limits_{A}^{N} N \right) \Big/ V_1$

$\langle 8 \rangle$ number density (of molecules or particles); density (of particles)
$\langle 9 \rangle$ Particularly in particle physics, no clear distinction is made between this kind-of-quantity and number concentration of component B, both being termed density.
$\langle 10 \rangle$ Sem—Spermatozoa; volumic num.(proc.) $= 2.1 \times 10^9$ L^{-1}

9.6.2 $\langle 2 \rangle$ **number concentration** of component B in system 1
$\langle 4 \rangle$ num.c.
$\langle 5 \rangle$ C_B
$\langle 6 \rangle$ number of entities of a component, N_B, divided by the volume of the system, V_1
$\langle 7 \rangle$ $C_B = N_B/V_1$
$\langle 8 \rangle$ molecular concentration; particle concentration; entity concentration
$\langle 9 \rangle$ The term *particle concentration* proved too restrictive, in that the kind-of-quantity can relate to any type of entity (§4.3). The same objection applies to the term molecular concentration. Both terms are also sometimes used for substance concentration (§9.88.1).
$\langle 10 \rangle$ B—Thrombocytes; num.c. $= 210 \times 10^9$ L^{-1}
$\langle 10 \rangle$ U(catheter)—Bacterium, nitrite producing; num.c.(proc.) $= 2 \times 10^9$ L^{-1}
$\langle 10 \rangle$ U—Crystals; num.c.(proc.) $= 32 \times 10^6$ L^{-1}
$\langle 10 \rangle$ P—Human immunodeficiency virus 1(RNA); num.c.(proc.) $= 20 \times 10^3$ L^{-1}
$\langle 10 \rangle$ Sem—Spermatozoa; num.c. $= 36 \times 10^9$ L^{-1}

9.6U   **Units of dimension L^{-3}**
$\langle 12 \rangle$ *Coherent SI units* (one) per cubic metre or reciprocal metre(s) or metre(s) to the power minus three (m^{-3})

Term for unit	Symbol of unit		Value in coherent SI unit
	Recommended	Other	
per picolitre	pL^{-1}, pl^{-1}		$= 10^{15}$ m^{-3}
per nanolitre	nL^{-1}, nl^{-1}		$= 10^{12}$ m^{-3}

(*Continued*)

Term for unit	Symbol of unit		Value in coherent SI unit
	Recommended	Other	
per cubic millimetre		mm^{-3}	$= 10^9 \ m^{-3}$
per microlitre	$\mu L^{-1}, \mu l^{-1}$		$= 10^9 \ m^{-3}$
per cubic centimetre		cm^{-3}	$= 10^6 \ m^{-3}$
per millilitre	mL^{-1}, ml^{-1}		$= 10^6 \ m^{-3}$
per cubic decimetre		dm^{-3}	$= 1000 \ m^{-3}$
per litre	L^{-1}, l^{-1}		$= 1000 \ m^{-3}$
per cubic metre	m^{-3}		$= 1 \ m^{-3}$

⟨19⟩ 'per litre' may be expressed by /L or by L^{-1}. The presentation 'L^{-1}' is preferred in this Compendium. The presentation /L may be retained for reasons of simplicity in the practice of laboratories.

9.7 Dimension L^{-4}

9.7.1 ⟨2⟩ **number concentration gradient** of component B at position (x,y,z)
⟨5⟩ **grad** C_B, ∇C_B
⟨6⟩ differential change in local number concentration of a component, C_B, with position, (x,y,z), divided by change in position
⟨7⟩ **grad** $C_B = dC_B/d(x,y,z)$
⟨8⟩ concentration gradient
⟨9⟩ Number concentration gradient is a vector kind-of-quantity.

9.7U **Units of dimension L^{-4}**
⟨12⟩ *Coherent SI units* (one) per metre to the power minus four or metre(s) to the power minus four (m^{-4})

9.8 Dimension M

9.8.1 ⟨2⟩ **mass**
⟨5⟩ m
⟨8⟩ weight
⟨9⟩ (1) Mass is the coefficient of proportionality between the force applied to a body and its acceleration.
⟨9⟩ (2) In common usage, the **mass** of a body is termed *weight*. It can be measured on a pan balance in vacuum against reference masses (also termed weights) and is independent of local acceleration to gravity of free fall. In physics, *weight* is used for the **force due to gravity** (§9.39.1.1), which depends on the acceleration of free fall (§9.34.1.2).
⟨9⟩ (3) For chemically defined components, a more informative kind-of-quantity is amount-of-substance (§9.83.1).
⟨9⟩ (4) To avoid confusion of the symbol m for mass or molality, the symbol b is given by ISO/IEC 80000-9 [4] for molality and is so used in this Compendium.
⟨10⟩ Pt—Liver; mass $= 1.6$ kg
⟨10⟩ Pt—Faeces; mass(24 h) $= 1.3$ kg

9.8.1.1 ⟨2⟩ **mass** of entity B
⟨3⟩ mass of individual; mass of particle; **rest mass**
⟨5⟩ M_B
⟨8⟩ liveweight; bodyweight; particle mass; nuclidic mass
⟨9⟩ (1) For elementary particles, the term **rest mass** refers to the mass of the particle in ground state, *i.e.* in neutral unexcited and uncombined state.

⟨9⟩ (2) **Rest mass of an atom** (of a nuclide), $_A$B, (or **nuclidic mass**), $m_{at,B}$ or $m_{a,B}$, is commonly termed *atomic mass*, which term also applies to the property of a multinuclidic element.

⟨10⟩ Electron; rest mass $= 0.910\,938\,291 \times 10^{-30}$ kg; $u_c = 0.000\,000\,040 \times 10^{-30}$ kg

⟨10⟩ Proton; rest mass $= 1.672\,621\,777$ yg; $u_c = 0.000\,000\,074$ yg

⟨10⟩ Neutron; rest mass $= 1.674\,927\,351$ yg; $u_c = 0.000\,000\,074$ yg

⟨10⟩ Atom of carbon-12; rest mass $= m(^{12}C) = 12$; $m_u \approx 19.926\,48$ yg where m_u is unified atomic mass constant (9.8U).

⟨10⟩ Atom of hydrogen-1; rest mass $= 1.673\,534\,0$ yg; $u_c = 0.000\,001\,0$ yg

⟨10⟩ U—Calculus; mass $= 2.3$ g

⟨10⟩ Pt(Arm; right)—Sweat; mass(iontophoresis) $= 55$ mg

9.8.1.2 ⟨2⟩ **mass** of system 1 or component B **given**

⟨5⟩ Δm

⟨8⟩ dose; Δ

⟨9⟩ *Dose* and *dosage* are commonly confused. In careful usage, *dose* is reserved for amount given (number, mass, volume or amount-of-substance) and *dosage* for dose with respect to, for instance, mass of patient (mass ratio §8.4.2), or to time, also termed dose rate. Where the chemical entity is known, expression as amount-of-substance (§9.83.1), substance rate (§9.94.1), or substance content rate (§9.98.1) are preferable.

⟨10⟩ Pt—Body; mass(proc.) $= 70$ kg

⟨10⟩ Pt—Lithium carbonate(administered); mass(p.o.) $= 250$ mg

⟨10⟩ Pt—Protirelin(administered); mass(i.v.) $= 250$ µg

9.8.1.3 ⟨2⟩ **mass** of component B **for effect E on half of population** (or sample 1 of biological system 1)

⟨5⟩ $m_{B,E,0.5}$, $m_{B,E}(\delta = 0.5)$

⟨6⟩ statistically derived mass of a component expected to produce a specified effect in half the organisms in that population or to produce a half-maximum effect in that biological system

⟨8⟩ median effective dose; effective dosage (at 50% effect); ED_{50}

⟨9⟩ This kind-of-quantity represents only one value of a function of mass of a biologically active component or material with respect to number fraction δ, of individuals in which a specified effect is detectable within a specified time. The function may be symbolized $m_{B,E}(\delta)$ and is commonly plotted on a logarithmic scale. A switch from mass to amount-of-substance of component is recommended (§7.1.2).

9.8.1.4 ⟨2⟩ **mass** of component B **for death of half of population L** (or sample) in stated time under stated environmental conditions

⟨5⟩ $m_{B,L,0.5}$, $m_{B,L}(\delta = 0.5)$

⟨6⟩ statistically derived mass of that component expected to kill half the organisms in that population

⟨8⟩ median lethal dose; LD_{50}

⟨9⟩ See §9.8.1.3.

9.8.1.5 ⟨2⟩ **mass of food** or nutrient B **ingested**

⟨5⟩ I_B

⟨8⟩ intake of food or nutrient

⟨9⟩ Food intake can better be expressed as a mean mass rate. Intake of nutrients can be expressed as mean substance rates (§9.94.1).

⟨10⟩ Pt—Fat(ingested); mass(1 d) $= 110$ g

9.8.2 ⟨2⟩ **entitic mass** of entities of component B in system 1

⟨5⟩ m_B/N_B

⟨6⟩ mass of a component divided by the number of specified entities of the component

9.8.2.1 ⟨2⟩ **entitic mass of atoms** of element B
⟨3⟩ **atomic mass**
⟨5⟩ $m_{at,B}$, $m_B/N_{at,B}$
⟨6⟩ rest mass of an element in a system, m'_B, in its nuclear and atomic ground state divided by the number of atoms of that element, $N_{at,B}$
⟨7⟩ $m_B = m'_B/N_{at,B}$
⟨9⟩ (1) The term *atomic mass* in this sense assumes typical terrestrial nuclidic composition of the element and is only a full synonym under that assumption. The term may also be used for specified atypical terrestrial systems or non-terrestrial systems.
⟨9⟩ (2) Present recommendations do not distinguish a symbol for entitic mass of element B from that for mass of a system (*e.g.* a sample) of element B.
⟨9⟩ (3) For a mononuclidic element (*i.e.* when all entities are equal in mass), the value of entitic mass of atoms equals the value of mass of an atom.
⟨9⟩ (4) $m_B = M_{r,B} \, m_u$

9.8.2.2 ⟨2⟩ **entitic mass of molecules** of compound B
⟨3⟩ **molecular mass**
⟨6⟩ rest mass of molecules of component B in a given system in its nuclear and molecular ground state divided by the number of molecules of that component
⟨7⟩ m_B/N_B
⟨8⟩ mass of molecule
⟨9⟩ Molecular mass in this sense represents the average mass of a group of molecules and is related to relative molar mass, $M_{r,B}$, by the atomic mass constant, m_u (§9.8U ⟨19⟩ (2)).
$m_B/N_B = M_{r,B} \, m_u$
⟨10⟩ Ercs(B)—Haemoglobin; entitic mass $= 32$ pg

9.8.3 ⟨2⟩ **mass carried** by a structure 1
⟨5⟩ m
⟨8⟩ load
⟨9⟩ 'Load' may alternatively be expressed as a force (§9.39.1) or as a normal stress.

9.8U **Units of dimension M**
⟨12⟩ *SI base unit* kilogram(s) (kg)
⟨16⟩ The kilogram is the unit of mass: it is equal to the mass of the international prototype of the kilogram (3rd CGPM 1901). Alternative definitions of the kilogram are under discussion (CIPM, CCU).

Term for unit	Symbol of unit		Value in coherent SI unit
	Recommended	Other	
megagram(s)	Mg		$= 1000$ kg
kilogram(s)	kg	Kg	$= 1$ kg
gram(s)	g	gr, gm	$= 0.001$ kg
milligram(s)	mg		$= 10^{-6}$ kg
microgram(s)	μg		$= 10^{-9}$ kg
gamma(s)		γ	$= 10^{-9}$ kg
nanogram(s)	ng		$= 10^{-12}$ kg
picogram(s)	pg		$= 10^{-15}$ kg
femtogram(s)	fg		$= 10^{-18}$ kg
attogram(s)	ag		$= 10^{-21}$ kg
unified atomic	m_u		$\approx 1.660\,538\,921 \times 10^{-27}$ kg
mass constant			$u_c = 0.000\,000\,073 \times 10^{-27}$ kg
unified atomic mass unit(s)	u	amu	$\approx 1.660\,538\,921 \times 10^{-27}$ kg
			$u_c = 0.000\,000\,073 \times 10^{-27}$ kg
dalton(s)	Da		$\approx 1.660\,538\,921 \times 10^{-27}$ kg
			$u_c = 0.000\,000\,073 \times 10^{-27}$ kg

⟨19⟩ (1) The spelling kilogram has been agreed internationally for the English language (CGPM).[2]

⟨19⟩ (2) See Table 5.4 on the (unified) atomic mass unit, also termed dalton. Numeric values in that unit are equal to those of molar mass in grams per mole. As a quantity, it is termed the (unified) atomic mass constant, m_u. It equals a twelfth of the rest mass of a neutral atom of the nuclide carbon-12 in its ground state.

⟨19⟩ (3) D should not be used as a symbol for dalton.

9.9 Dimension M^{-1}

⟨2⟩ **number content** of component B in system 1
⟨4⟩ num.cont.
⟨5⟩ N_B/m_1
⟨6⟩ number of a specified entity or group of entities (B) divided by the mass of system 1
⟨8⟩ count
⟨10⟩ Faes—*Oxyuris vermicularis* egg; num.cont. $= 500$ kg^{-1}

9.9U Units of dimension M^{-1}
⟨12⟩ *Coherent SI units* (one) per kilogram or reciprocal kilogram(s) (kg^{-1})

9.10 Dimension $L^{-1}M$

9.10.1 ⟨2⟩ **lineic mass** of system 1 or of a component B in a system 1
⟨5⟩ m_l, ρ_l
⟨6⟩ mass of a body, m_1, for instance a fibre or of a component, m_B, divided by its length, l
⟨7⟩ $m_{l,1} = m_1/l_1$
⟨8⟩ linear density

9.10U Unit of dimension $L^{-1}M$
⟨12⟩ *Coherent SI unit* kilogram(s) per metre (kg m^{-1})

9.11 Dimension L^2M

9.11.1 ⟨2⟩ **moment of inertia** of a body about an axis
⟨3⟩ **dynamic moment of inertia**
⟨5⟩ I, J
⟨6⟩ sum or integral of the products of the mass elements of a body, m_i, and the squares of their respective distances from the axis, r_i
⟨7⟩ $I = \int_1^n (m_i r_i^2)$

9.11U Unit of dimension L^2M
⟨12⟩ *Coherent SI unit* kilogram(s) metre squared (kg m^2)

9.12 Dimension $L^{-2}M$

9.12.1 ⟨2⟩ **areic mass** of system 1 for specified surface or interface
⟨5⟩ m_A, ρ_A
⟨6⟩ mass of a system, m_1, divided by a specified area of the system, A_1
⟨7⟩ $m_A = m_1/A_1$
⟨8⟩ surface density; ρ_S
⟨19⟩ The terms "Body Mass Index" (BMI) and "Quetelet Index" relate to 'mass of person divided by the square of body height' and is therefore not 'areic mass', although of the same measurement dimension and measurement unit. The term "mass parameter" would be preferable.

9.12U **Units of dimension $L^{-2}M$**
⟨12⟩ *Coherent SI unit* kilogram(s) per square metre (kg m^{-2})

Term for unit	Symbol of unit Recommended	Value in coherent SI unit
kilogram(s) per square metre	kg m^{-2}	$= 1$ kg m^{-2}
gram(s) per square metre	g m^{-2}	$= 0.001$ kg m^{-2}
milligram(s) per square metre	mg m^{-2}	$= 10^{-6}$ kg m^{-2}

9.13 Dimension L^2M^{-1}

9.13.1 ⟨2⟩ **massic area** of system 1 for specified surface or interface
⟨5⟩ a_1
⟨6⟩ area of a system, A_1, divided by the mass of the system, m_1
⟨7⟩ $a_1 = A_1/m_1$
⟨8⟩ specific area

9.13.2 ⟨2⟩ **massic area rotance** of optically active component B in solution 1 to linearly polarized light λ
⟨3⟩ **massic optical rotatory power**
⟨5⟩ $\alpha_{\rho,B}$, $\alpha_{m,B}$
⟨6⟩ optical rotation (or rotance) of polarized light by the solution, α, divided by the path length of the light, l, and by the mass concentration of the component, ρ_B
⟨7⟩ $\alpha_{\rho,B} = \alpha/(l\,\rho_B)$
⟨8⟩ specific optical rotatory power
⟨9⟩ Molar area rotance (§9.86.2) is preferable.

9.13.3 ⟨2⟩ **napierian attenuation coefficient** to specific radiation λ by system 1
⟨3⟩ specific napierian extinction coefficient
⟨5⟩ $\alpha_{s,B}$, $\alpha_{s,1}$
⟨6⟩ attenuation of a radiation of a system, $-\delta P/P$, divided by path length of the radiation, l, and by the mass concentration of an attenuating component, ρ_B, or by the volumic mass of the system, ρ_1
⟨7⟩ (1) $\alpha_{s,B} = (d\ln P/dl)/\rho_B$
⟨7⟩ (2) $\alpha_{s,1} = (d\ln P/dl)/\rho_1$
⟨8⟩ attenuation coefficient; extinction coefficient; mass extinction coefficient

9.13.4 ⟨2⟩ **absorption coefficient** to radiation λ of absorbing component B
⟨6⟩ logarithmically decreasing radiation due to an absorbing component, $-\delta P/P$, divided by path length of the radiation, l, and by mass concentration of the component, ρ_B
⟨7⟩ $\alpha_{s,B} = (d\ln P/dl)/\rho_B$
⟨8⟩ absorptivity; specific absorption coefficient; specific absorptivity
⟨9⟩ (1) Analogous coefficients in terms of decadic logarithms are commonly used in optical spectroscopy. As long as the reduction in the radiation is due only to absorption, the coefficients are termed **absorption coefficients**.
⟨9⟩ (2) Molar area absorbance is preferred (§7.3.6 and §9.86.3).

9.13U **Units of dimension L^2M^{-1}**
⟨12⟩ *Coherent SI unit of massic area* square metre(s) per kilogram (m^2 kg^{-1})
⟨12⟩ *Coherent SI units of massic area rotance* radian square metre(s) per kilogram or square metre(s) per kilogram (rad m^2 kg^{-1} = (m^2 kg^{-1}))

9.14   Dimension $L^{-3}M$

9.14.1 $\langle 2 \rangle$ **volumic mass** of system 1
$\langle 3 \rangle$ **mass density**
$\langle 5 \rangle$ ρ_1
$\langle 6 \rangle$ mass of the system, m_1, divided by its volume, V_1
$\langle 7 \rangle$ $\rho_1 = m_1/V_1$
$\langle 8 \rangle$ density; specific gravity; specific weight
$\langle 9 \rangle$ (1) Volumic mass depends on temperature and, for gases, on pressure, which form an integral part of the specification of any volumic mass. For liquids, ambient pressure of about 100 kPa is assumed unless otherwise stated.
$\langle 9 \rangle$ (2) The term *specific weight* was formerly used for *weight density* (*i.e.* volumic force due to gravity, SI unit N m^{-3}, dimension $L^{-2}MT^{-2}$) but sometimes also for volumic mass.
$\langle 9 \rangle$ (3) The term *density* is ambiguous.
$\langle 10 \rangle$ Water—; volumic mass $(t = 4\ °C) = 999.972$ g L^{-1}
$\langle 10 \rangle$ U—; volumic mass $(t = 20\ °C) = 1.216$ kg L^{-1}

9.14.1.1 $\langle 2 \rangle$ **partial volumic mass** of component B in system 1
$\langle 3 \rangle$ **partial mass density**
$\langle 5 \rangle$ ρ_B
$\langle 6 \rangle$ differential change in mass of a system when a component enters or leaves the system divided by the resulting change in volume of the system
$\langle 7 \rangle$ $\rho_B = \delta m_B/\delta V_1$
$\langle 8 \rangle$ partial density

9.14.2 $\langle 2 \rangle$ **mass concentration** of component B in system 1
$\langle 4 \rangle$ mass c.
$\langle 5 \rangle$ ρ_B, γ_B
$\langle 6 \rangle$ mass of component, m_B, divided by volume of the system, V_1
$\langle 7 \rangle$ $\rho_B = m_B/V_1$
$\langle 8 \rangle$ concentration (§7.1.2, Table 7.3); mass density; c_B
$\langle 9 \rangle$ (1) Where an elementary entity can be defined, substance concentration (§9.88.1) is preferable, because it gives more biochemical information.
$\langle 9 \rangle$ (2) Mass concentration varies with temperature and, especially for gases, pressure of the mixture. Contrast mass fraction (§8.4.1).
$\langle 9 \rangle$ (3) Mass concentration of pure solvent, ρ_{A^*}, is the same as volumic mass of pure solvent, $\rho_1(A^*)$, where the star indicates 'pure'.
$\langle 10 \rangle$ P—Protein; mass c. $= 40$ g L^{-1}
$\langle 10 \rangle$ P—Vitamin D-binding protein; mass c. $= 215$ mg L^{-1}
$\langle 10 \rangle$ P—Immunoglobulin A; mass c.(RM-DA470K; proc.) $= 1.84$ g L^{-1}
$\langle 10 \rangle$ P—Somatotropin; mass c.(IS 98/574; proc.) $= 4.1$ µg L^{-1}
$\langle 10 \rangle$ P—Ferritin; mass c.(IS 80/602; proc.) $= 90$ µg L^{-1}
$\langle 10 \rangle$ Air(expired)—Ethanol; mass c.(breath analyser) $= 20$ mg L^{-1}
This measuring system is used in legal metrology in some geographical areas (Switzerland, Australia,...).

9.14U   **Units of dimension $L^{-3}M$**
$\langle 12 \rangle$ *Coherent SI unit* kilogram(s) per cubic metre (kg m^{-3})

Term for unit	Symbol of unit		Value in coherent SI unit
	Recommended	Other	
megagram(s) per cubic metre	Mg m^{-3}		$= 1000$ kg m^{-3}
kilogram(s) per litre	kg L^{-1}, kg l^{-1}		$= 1000$ kg m^{-3}
gram(s) per millilitre		g mL^{-1}, g/ml	$= 1000$ kg m^{-3}

(*Continued*)

Term for unit	Symbol of unit		Value in coherent SI unit
	Recommended	Other	
gram(s) per decilitre		g/dL, g/dl	$= 10$ kg m^{-3}
kilogram(s) per hectolitre		kg/hL, kg/hl	$= 10$ kg m^{-3}
per cent	%		$= 10$ kg m^{-3}
kilogram(s) per cubic metre	kg m^{-3}		$= 1$ kg m^{-3}
gram(s) per litre	g L^{-1}, g l^{-1}		$= 1$ kg m^{-3}
per mille	‰		$= 1$ kg m^{-3}
gram(s) per cubic metre	g m^{-3}		$= 0.001$ kg m^{-3}
milligram(s) per litre	mg L^{-1}, mg l^{-1}		$= 0.001$ kg m^{-3}
microgram(s) per decilitre		µg/dL, µg/dl	$= 10 \times 10^{-6}$ kg m^{-3}
milligram(s) per cubic metre	mg m^{-3}		$= 10^{-6}$ kg m^{-3}
microgram(s) per litre	µg L^{-1}, µg l^{-1}		$= 10^{-6}$ kg m^{-3}
microgram(s) per cubic metre	µg m^{-3}		$= 10^{-9}$ kg m^{-3}
nanogram(s) per litre	ng L^{-1}, ng l^{-1}		$= 10^{-9}$ kg m^{-3}

⟨19⟩ (1) In clinical chemistry, the litre is preferred to the cubic metre (§5.10.2).
⟨19⟩ (2) ISO/IEC 80000-1[5] and CGPM[2] deprecate expressions such as % (w/v) and ‰ (w/v).

9.15 Dimension L³M⁻¹

9.15.1   ⟨2⟩ **massic volume** of system 1
⟨5⟩ v_1
⟨6⟩ volume of a system, V_1, divided by its mass, m_1
⟨7⟩ $v_1 = V_1/m_1$
⟨8⟩ specific volume
⟨9⟩ (1) Massic volume is the reciprocal of volumic mass, ρ_1:

$$v_1 = 1/\rho_1$$

⟨9⟩ (2) Massic volume is used in estimating the molar mass of macromolecules (*e.g.* nucleic acids) and colloidal particles (*e.g.* viruses) by means of the sedimentation coefficient (§9.17.3.1).

9.15.1.1   ⟨2⟩ **partial massic volume** of component B in system 1
⟨5⟩ v_B
⟨6⟩ differential change in volume of a system, V_1, when a component is added divided by the mass of the component added, m_1
⟨7⟩ $v_B = dV_1/dm_B$
⟨8⟩ partial specific volume
⟨9⟩ It is necessary to specify temperature and pressure for this kind-of-quantity.

9.15.2   ⟨2⟩ **volume content** of component B
⟨4⟩ vol.cont.
⟨5⟩ v_B
⟨6⟩ volume of an (isolated) component, V_B, divided by the mass of the system, m_1
⟨7⟩ $v_B = V_B/m_1$

⟨8⟩ content; specific volume content

⟨9⟩ (1) Volume content depends on temperature and, especially for gases, the pressure of the component.

⟨9⟩ (2) For chemical components, substance content is usually more informative.

⟨9⟩ (3) The compositional quantity, volume content, is distinguished in term from the material property massic volume (§9.15.1).

⟨10⟩ Liquid solvent—Gas dissolved (saturated; corr. for temp. and pr.); vol.cont.

⟨10⟩ Pt–Blood; vol.cont. $= 70$ mL kg^{-1}

9.15U **Units of dimension L^{3}M^{-1}**

⟨12⟩ *Coherent SI unit* cubic metre(s) per kilogram (m^3 kg^{-1})

Term for unit	Symbol of unit Recommended	Other	Value in coherent SI unit
cubic metre(s) per kilogram	m^3 kg^{-1}		$= 1$ m^3 kg^{-1}
litre(s) per kilogram	L kg^{-1}, l kg^{-1}		$= 10^{-3}$ m^3 kg^{-1}
cubic centimetre(s) per gram		cm^3/g	$= 10^{-3}$ m^3 kg^{-1}
millilitre(s) per gram		mL/g, ml/g	$= 10^{-3}$ m^3 kg^{-1}
millilitre(s) per kilogram	mL kg^{-1}, ml kg^{-1}		$= 10^{-6}$ m^3 kg^{-1}
per mille	‰		$= 10^{-6}$ m^3 kg^{-1}
per cent	%		$= 0.01 \times 10^{-3}$ m^3 kg^{-1}
cubic millimetre(s) per kilogram	mm^3 kg^{-1}		$= 10^{-9}$ m^3 kg^{-1}
microlitre(s) per kilogram	μL kg^{-1}, μl kg^{-1}		$= 10^{-9}$ m^3 kg^{-1}

⟨19⟩ ISO/IEC 80000-1[5] deprecates expressions such as % (v/w) and ‰ (v/w).

9.16 Dimension L^{-4}M

9.16.1 ⟨2⟩ **volumic mass gradient** of system 1 at position (x,y,z)

⟨5⟩ **grad** ρ_1, $\nabla \rho_1$

⟨6⟩ differential change in local volumic mass of a system, ρ_1, with position, (x,y,z), divided by change in position

⟨7⟩ **grad** $\rho_1 = \mathrm{d}\rho_{1,x,y,z}/\mathrm{d}(x,y,z)$

⟨8⟩ mass density gradient

⟨9⟩ (1) Volumic mass gradient is a vector kind-of-quantity.

⟨9⟩ (2) Colloidal components may be fractionated by centrifuging in a fluid with a gradient obtained with a suitable solute, for instance potassium bromide in water.

9.16.2 ⟨2⟩ **mass concentration gradient** of component B at position (x,y,z)

⟨4⟩ mass c. gradient

⟨5⟩ **grad** ρ_{B}, $\nabla \rho_{\mathrm{B}}$

⟨6⟩ differential change in local mass concentration of a component, ρ_{B}, with position, (x,y,z), divided by the change in position

⟨7⟩ $\nabla \rho_{\mathrm{B}} = \mathrm{d}\rho_{\mathrm{B},x,y,z}/\mathrm{d}(x,y,z)$

⟨8⟩ concentration gradient

⟨9⟩ Mass concentration gradient is a vector kind-of-quantity.

9.16U **Unit of dimension L^{-4}M**
⟨12⟩ *Coherent SI unit* kilogram(s) per metre to the power four (kg m^{-4})

9.17 Dimension T

9.17.1 ⟨2⟩ **time**
⟨3⟩ **duration; time elapsed; down time** of instrument 1; **infectious time** (or *infectious period*), t_i; **latent time** (or *latent period*) of infection B in organism 1; **life expectancy** of organism 1, $E(t)$, e_t; **age** of organism 1; lifetime; **time of consumption** of total prepared sample, t_{tot}; **minimum consumption time** of prepared sample, t_{min}; **minimum time of measurement** of operation B with system 1; operational time; **practical response time** of measuring instrument or system 1, b_1; **resolving time; travel time** (chromatography), t_{tv}; **travel time; turn-over time; transit time,** t_{ts}
⟨5⟩ t
⟨8⟩ period; time period
⟨9⟩ (1) The designation **time difference** (Δt) may be used instead of time.
⟨10⟩ Skin—Arteriolar bleeding; time difference(template) = 180 s
⟨9⟩ (2) *Time* may also mean a specific moment in a given calendar, which may be termed **date** (or **calendar time**) and **time of day** (or clock-time) and should be expressed in accordance with ISO 2014[6] and ISO 8601.[7] Dates may also be represented partly in words, 2016-04-16 T14:30:30 or 2016-04-16 14h30m30s or 16 April 2016 at 14:30:30.
⟨9⟩ (3) Many measured quantities are a function of time and may also be influenced by a previous event at a specified time. Both may need to be specified (§6.8.1–2).
⟨9⟩ (4) When a mean rate is reported, the limits of the time interval over which measurements are made needs to be stated (§6.8.3–4).
The property being studied might have a biological rhythm, then clock-times should be specified.
⟨10⟩ B—Coagulation; surface-induced; time(proc.) = 12 s
⟨10⟩ Pt(B)—Lymphocytes; doubling time(proc.) = 13 mth
⟨10⟩ Pt—Amenorrhoea; duration = 4 wk
⟨10⟩ Pt—Sexual abstinence; time = 8 d
⟨10⟩ Pt—Dexamethasone(administered); duration(between doses) = 9 h
⟨10⟩ Pt—Urine sampling; duration = 1 d
⟨10⟩ Pt—Capillary bleeding; time(Ivy) = 210 s
⟨10⟩ Sem—Sample transport; time(proc.) = 80 min

9.17.1.1 ⟨2⟩ **time of centrifugation**
⟨5⟩ t
⟨6⟩ time difference from switching on until switching off a centrifuge
⟨9⟩ The time for deceleration is not included.

9.17.2 ⟨2⟩ **entitic time** of regular phenomenon B
⟨5⟩ T
⟨6⟩ time during which a regular event occurs, Δt, divided by the number of repetitions of that event, ΔN_B
⟨7⟩ $T = \Delta t / \Delta N_B$
⟨8⟩ period; periodic time

9.17.3 ⟨2⟩ **time coefficient**
⟨5⟩ τ, k
⟨6⟩ (1) time after which the quantity would reach its limit if it maintained its initial rate of change
⟨7⟩ (1) $\tau = \mathrm{d}t / |\mathrm{d}\ln x|$

⟨7⟩ (2) If a quantity is a function of time given by

$$F(t) = A + B \exp(-1/\tau)$$

then τ is the time coefficient

⟨8⟩ characteristic time (interval); relaxation time; time constant; T

⟨9⟩ Use of time coefficient for compositional changes in kinds-of-quantity (*e.g.* mass fraction and substance concentration) assumes that the size of the system changes only negligibly.

9.17.3.1 ⟨2⟩ **sedimentation coefficient** of suspended component B in fluid system 1

⟨5⟩ s_B

⟨6⟩ (1) reciprocal of the sedimentation coefficient, k_B, of the component passing a given position in the direction of gravitational or centrifugal acceleration

⟨7⟩ $s_B = k_B^{-1}$
$$= -N_B \, dt/dN_B$$
$$= -dt/d\ln N_B$$

⟨6⟩ (2) product of the velocity of sedimentation and acceleration

⟨8⟩ sedimentation velocity

⟨9⟩ In current usage, subscripts are added to the symbol to indicate temperature and medium, and superscripts to indicate concentration.

9.17.3.2 ⟨2⟩ **half-life** of component B subject to a specified process in system 1

⟨3⟩ half-value time; half-time; half-disappearance time

⟨5⟩ $T_{0.5}$, $T_{1/2}$, $t_{1/2}$

⟨6⟩ time taken for the amount (*e.g.* number, volume, mass, amount-of-substance or absorbed dose) to reach the arithmetic mean of its initial and final equilibrium value, assuming a first-order process

⟨9⟩ (1) For a reactant B subject to higher-order kinetics, half-time is proportional to $c(A)_\alpha^{1-\alpha}$ where α is the partial order with respect to B; it is therefore independent of substance concentration c, only when the partial order is 1.

⟨9⟩ (2) For radioactive (exponential) decay, it is related to decay coefficient, λ (§9.18.4.5), and to time coefficient, τ (§9.17.3), by:

$$T_{1/2} = (\ln 2)/\lambda = \tau \ln 2$$

For first-order (or pseudo-first-order) chemical reactions, it is related to rate coefficient k, by:

$$t_{1/2} = (\ln 2)/k$$

⟨9⟩ (3) The concept applies to higher-order reactions only if the concentrations of the reactants are in their stoichiometric ratios. If the kinetic equation is more complicated, there is no simple relationship between the half-life of a reactant and the rate coefficients.

⟨9⟩ (4) If a process cannot be defined exponentially, the process should be described by a series of values at different times.

⟨9⟩ (5) The quantity that is halved may be an amount (*e.g.* number, volume, mass, amount-of-substance) or a compositional quantity (*e.g.* mass fraction, substance concentration).

⟨10⟩ Pt(B)—Phenytoin elimination; half-life(oral dose; $n = 0.4$ mmol) = 75 ks

⟨10⟩ Pt(B)—Erythrocyte elimination; half-life(proc.) = 32 d

9.17U **Units of dimension T**

⟨12⟩ *SI base unit* second(s)

⟨16⟩ The second is the duration of 9 192 631 770 periods of the radiation corresponding to the transition between the two hyperfine levels of the ground state of the caesium-133 atom (13th CGPM 1967, Resolution 1). This definition refers to a caesium atom at rest at a thermodynamic temperature of 0 K (CIPM 1997).

In this definition, it is intended that the caesium atom at a $T = 0$ K is unperturbed by black-body radiation. The frequency of primary standards should therefore be corrected for the frequency shift due to the ambient radiation, as stated by the CCTF (1999). See 9.1U $\langle 16 \rangle$.

Term for unit	Symbol of unit		Value in coherent SI unit
	Recommended	Other	
gigasecond(s)	Gs		$= 10^9$ s
year(s) (tropical)	a	year, yr	$\approx 31.556\,93 \times 10^6$ s
month(s)		mth	$= 30.4$ d $= 2.626\,56 \times 10^6$ s
megasecond(s)	Ms		$= 10^6$ s
week(s)		wk	$= 604.800 \times 10^3$ s
day(s)	d	da	$= 86.400 \times 10^3$ s
hour(s)	h	hr	$= 3600$ s
kilosecond(s)	ks		$= 1000$ s
minute(s)	min	mn, m	$= 60$ s
second(s)	s	sec	$= 1$ s
second(s) per neper		s Np^{-1}	$= 1$ s
millisecond(s)	ms		$= 10^{-3}$ s
microsecond(s)	μs		$= 10^{-6}$ s
nanosecond(s)	ns		$= 10^{-9}$ s
picosecond(s)	ps		$= 10^{-12}$ s
femtosecond(s)	fs		$= 10^{-15}$ s
atomic unit(s) of time	$\hbar/E_h$		$\approx 24.188\,843\,265\,05 \times 10^{-18}$ s
			$u_c = 0.000\,000\,000\,16 \times 10^{-18}$ s
attosecond(s)	as		$= 10^{-18}$ s

$\langle 19 \rangle$ (1) The second was originally defined as the fraction $1/86\,400$ of the mean solar day, the definition of day being left to astronomers. In order to define the unit of time more precisely, the **tropical year** was adopted as standard in 1960. The atomic standard of time, based on energy transitions in atoms or molecules, can be reproduced much more accurately, a few parts in 10^{14}, and was adopted in 1967 (13th CGPM). The time definition is being changed (CCU 2011).

$\langle 19 \rangle$ (2) Customary units of time, based on the 12-digit system, lead to difficulties of calculation and of comparison. In expression of derived quantities, results should, whenever possible, be converted to the second, in order to retain the advantages of coherent units. Minutes, hours and days are recognized for use with SI (§5.10.3, Table 5.4), because of their importance and widespread use, but CQUCC recommends limitation of their use in compound units and restriction in a single set of data to a single unit of time, in order to simplify comparisons. In biological phenomena, the **day**, *week*, and the variables *month* and *year* are sometimes relevant.

$\langle 19 \rangle$ (3) The symbol m for minute should be avoided because of a possible confusion with the recommended symbol for metre.

$\langle 19 \rangle$ (4) The symbol "a" for year is suggested by ISO/IEC 80000-3.[8]

$\langle 19 \rangle$ (5) The *Svedberg unit* (S, Sv), formerly used for sedimentation coefficient, is not recommended.

9.18   Dimension T^{-1}

9.18.1 $\langle 2 \rangle$ **number rate** of entities of component B in system 1
$\langle 4 \rangle$ num.rate
$\langle 5 \rangle$ f, ν, Φ_N, $\dot{N}$
$\langle 6 \rangle$ number of entities of a component, N_B, during a time, t, divided by that time

⟨7⟩ $f = dN_B/dt$

⟨10⟩ Pt(U)—Erythrocytes; num.rate(10:00/13:00) $= 6 \times 10^3\,\mathrm{min}^{-1}$

9.18.1.1 ⟨2⟩**number rate of photons** λ

⟨5⟩ $\Phi_{N,\Phi}$

⟨6⟩ number of photons, N_Φ, in a time t, divided by that time

⟨7⟩ $\Phi_{N,\Phi} = dN_\Phi/dt$

⟨8⟩ photon flux; Φ_p, Φ

⟨9⟩ (1) The number rate of photons is related to the differential quotient of energy rate of the radiation to wavelength, $\Phi_{e,\lambda}$:

$$\Phi_{N,\Phi} = \int_0^1 \Phi_{e,\lambda}\lambda/h\,c\,d\lambda$$

⟨9⟩ (2) 'Photon flux' and its derivatives, *e.g.* 'photon intensity', $I_{N,\Phi}$, 'photon radiance', 'photon exitance', 'photon irradiance', 'photon exposure', can be expressed in less cumbersomely large numeric values as substance rate of photons and its derivatives instead of as number rate and its derivatives, by dividing values by the Avogadro constant.

9.18.1.2 ⟨2⟩ **number rate of regular event** B

⟨3⟩ **frequency**

⟨5⟩ f, ν

⟨6⟩ number of regular events, N_B, in a time, t, divided by that time

⟨7⟩ $f = N_B/t$

⟨9⟩ (1) Number rate of a periodic phenomenon is the reciprocal of its entitic time (or periodic time), T:

$$f = 1/T$$

⟨9⟩ (2) In contrast to wavelength and wavenumber, frequency of an electromagnetic radiation is independent of the medium.

⟨9⟩ (3) The frequency of electromagnetic radiation is proportional to photon energy, usually expressed as molar energy of photons, $E_{m,\Phi}$:

$$\nu = E_{m,\Phi}\,N_A/h$$

where N_A is the Avogadro constant and h is the Planck constant.

⟨9⟩ (4) Frequency is also related to the speed of the radiation in a given medium, c_1, and the wavelength in that medium, λ_1:

$$\nu = c_1/\lambda_1$$

⟨10⟩ Person—Cardiac pulsation; number rate $= 1.11\,\mathrm{s}^{-1} = 1.11\,\mathrm{Hz}$

9.18.1.3 ⟨2⟩ **number rate of rotation** of rotor

⟨3⟩ **rotational frequency**; rate of rotation; rate of revolution; centrifugal speed

⟨5⟩ f_{rot}

⟨6⟩ number of complete rotations (or revolutions), N_B, in a time, t, divided by that time

⟨7⟩ $f_{rot} = dN_{rot}/dt_{rot}$

⟨8⟩ n

⟨9⟩ The expression in number of rotations per min is not by itself informant because the value of centrifugal acceleration depends on diameter of the used rotor.

9.18.2 ⟨2⟩ **activity referred to a radionuclide** B in system 1

⟨5⟩ A_B

⟨8⟩ radionuclidic activity; radioactivity; activity

⟨10⟩ Pt—^{58}Co-Cobalamin(administered); radioact.(p.o.) $= 28\,\mathrm{kBq}$

9.18.3   ⟨2⟩ **angular velocity** of rotor
       ⟨3⟩ pulsatance; angular frequency; circular frequency
       ⟨5⟩ ω
       ⟨7⟩ (1) $\omega = d\varphi/dt$
       ⟨7⟩ (2) $\omega = 2\pi f$ where f is frequency.

9.18.4   ⟨2⟩ **rate coefficient** of exponentially changing quantity Q
       ⟨5⟩ δ, λ
       ⟨6⟩ relative change in an exponentially changing quantity, Q, as a differential to time
       ⟨7⟩ $\delta = |d\ln Q/dt| = |(dQ/dt)/Q|$
       ⟨8⟩ damping coefficient, rate constant
       ⟨9⟩ (1) The term damping coefficient applies where the time function is of amplitude.
       ⟨9⟩ (2) If the size of the system remains constant, a process can be described in terms of the change in a compositional kind-of-quantity (*e.g.* substance concentration). Otherwise an extensive kind-of-quantity (*e.g.* amount-of-substance of the component eliminated) must be used.

9.18.4.1   ⟨2⟩ **rate coefficient for equimolecular reaction** B
       ⟨5⟩ k
       ⟨6⟩ coefficient of proportionality in the equation for the rate of a chemical reaction
       ⟨8⟩ rate constant
       ⟨9⟩ (1) The coefficient in the equation for the rate of a chemical reaction may have a dimension more complex than T^{-1}, according to the form of the stoichiometric equation, *e.g.* $L^{3}T^{-1}N^{-1}$.
       ⟨9⟩ (2) When the dimension is T^{-1}, this kind-of-quantity is usually termed rate constant.
       ⟨9⟩ (3) It is necessary to specify temperature and pressure for this kind-of-quantity.

9.18.4.2   ⟨2⟩ **rate coefficient for sedimentation** of component B in fluid 1
       ⟨5⟩ k_B
       ⟨6⟩ number fraction of particles of a component passing a given position in the direction of an applied force divided by the time of passage
       ⟨7⟩ $k_{B,1} = -(dN_B/dt)/N_B$
             $= -d\ln N_B/dt$
       ⟨9⟩ The applied force may be from gravitational acceleration in column chromatography, rotational acceleration in a centrifuge, or electromotive force in electrophoresis.

9.18.4.3   ⟨2⟩ **rate coefficient for elimination** of component B from system 1
       ⟨3⟩ elimination rate coefficient
       ⟨5⟩ $k_{elim,B}$
       ⟨8⟩ elimination rate constant
       ⟨10⟩ Pt(B)—Acetaminophen; elimination rate coefficient(proc.) $= 0.27$ h^{-1}

9.18.4.4   ⟨2⟩ **decay coefficient for absorbed dose** of component B
       ⟨3⟩ **disintegration coefficient**
       ⟨5⟩ λ_B
       ⟨6⟩ number rate of atoms of a component decaying (or disintegrating), $-dN/dt$, divided by the number of atoms of that component, N
       ⟨7⟩ $\lambda = (-dN/dt)/N$
       ⟨8⟩ decay constant; disintegration constant

9.18.4.5 $\langle 2 \rangle$ **decay coefficient for absorbed dose** of component B **by process B → C**
$\langle 3 \rangle$ **partial decay coefficient**
$\langle 6 \rangle$ number rate of atoms of a component decaying (or disintegrating) by one of several modes of decay divided by the number of atoms of that component
$\langle 8 \rangle$ partial decay constant

9.18.5 $\langle 2 \rangle$ **relative mass rate** of system 1 to system 2
$\langle 8 \rangle$ dosage
$\langle 9 \rangle$ (1) See §9.8.1.3, $\langle 9 \rangle$.
$\langle 9 \rangle$ (2) As long as system 1 is much smaller than system 2, such a quantity can be considered as a mass fraction rate.

9.18.6 $\langle 2 \rangle$ **molar catalytic activity** of enzyme E with measuring system 1
$\langle 4 \rangle$ molar cat.act.
$\langle 5 \rangle$ $z_{m,E}$
$\langle 7 \rangle$ $z_{m,E} = z_E/n_E$ (Table 7.2)

9.18U **Units of dimension T^{-1}**
$\langle 12 \rangle$ *Coherent SI units of number rate* (one) *per second* or *reciprocal second(s)* (s^{-1})
$\langle 12 \rangle$ *Coherent SI unit of frequency of a regular periodic phenomenon hertz(es)* ($Hz = s^{-1}$)
$\langle 12 \rangle$ *Coherent SI unit of absorbed dose becquerel(s)* ($Bq = s^{-1}$)
$\langle 12 \rangle$ *Coherent SI unit of angular velocity radian(s) per second* ($rad\ s^{-1} = s^{-1}$)
$\langle 12 \rangle$ *Coherent units of molar catalytic activity katal(s) per mole* or *reciprocal second(s)* ($kat\ mol^{-1} = mol\ s^{-1}\ mol^{-1} = s^{-1}$)

Term for unit	Symbol of unit Recommended	Other	Value in coherent SI unit
per femtosecond	fs^{-1}		$= 10^{15}\ s^{-1}$
petahertz(es)	PHz		$= 10^{15}$ Hz
per picosecond	ps^{-1}		$= 10^{12}\ s^{-1}$
terahertz(es)	THz		$= 10^{12}$ Hz
curie(s)		Ci, c	$= 37 \times 10^9$ Bq
per nanosecond	ns^{-1}		$= 10^9\ s^{-1}$
gigahertz(es)	GHz		$= 10^9$ Hz
gigabecquerel(s)	GBq		$= 10^9$ Bq
millicurie(s)		mCi	$= 37 \times 10^6$ Bq
per microsecond	μs^{-1}		$= 10^6\ s^{-1}$
megahertz(es)	MHz		$= 10^6$ Hz
microcurie(s)		μCi	$= 37 \times 10^3$ Bq
per millisecond	ms^{-1}		$= 1000\ s^{-1}$
kilohertz(es)	kHz		$= 1000$ Hz
per second	s^{-1}		$= 1\ s^{-1}$
hertz(es)	Hz		$= 1$ Hz
becquerel(s)	Bq		$= 1$ Bq
revolution(s) per second		r.p.s., rev/s	$= 1\ s^{-1}$
neper(s) per second		Np/s	$= 1\ s^{-1}$
radian(s) per second	$rad\ s^{-1}$		$= 1\ s^{-1}$
katal(s) per mole	$kat\ mol^{-1}$		$= 1\ s^{-1}$
per minute		min^{-1}	$\approx 16.666\,67 \times 10^{-3}\ s^{-1}$
per kilosecond	ks^{-1}		$= 10^{-3}\ s^{-1}$
gram(s) per kilogram second	$g\ kg^{-1}\ s^{-1}$		$= 10^{-3}\ s^{-1}$

(*Continued*)

Term for unit	Symbol of unit		Value in coherent
	Recommended	Other	SI unit
per hour		h^{-1}	$\approx 0.277\,777\,8 \times 10^{-3}$ s^{-1}
per day	d^{-1}		$\approx 11.574\,07 \times 10^{-6}$ s^{-1}
per year	a^{-1}	year^{-1},	$\approx 31.688\,76 \times 10^{-9}$ s^{-1}
		yr^{-1}	
per annum		p.a.	$\approx 31.688\,76 \times 10^{-9}$ s^{-1}

$\langle 19 \rangle$ (1) The hertz (Hz) is permitted by CGPM only for periodic phenomena.
$\langle 19 \rangle$ (2) The neper per second is recognized by ISO/IEC 80000-1[5] for *damping coefficient*, but is not mentioned by BIPM (1991). Its definition is unclear.[9]
$\langle 19 \rangle$ (3) The becquerel is used only for the activity of a radionuclide.
$\langle 19 \rangle$ (4) See §9.17U, $\langle 19 \rangle$ (2) for considerations in choice of time units.

9.19 Dimension L^{-1}T

9.19.1 $\langle 2 \rangle$ **lineic time**
$\langle 7 \rangle$ dt/dl

9.19U **Units of dimension L^{-1}T**
$\langle 12 \rangle$ *Coherent SI unit* second(s) per metre (s m^{-1})

Term for unit	Symbol of unit		Value in coherent SI unit
	Recommended	Other	
megasecond(s) per metre	Ms m^{-1}		$= 10^6$ s m^{-1}
hour(s) per centimetre		h cm^{-1}	$= 360 \times 10^3$ s m^{-1}
day(s) per metre	d m^{-1}		$= 86.4 \times$ s m^{-1}
second(s) per centimetre		s cm^{-1}	$= 100$ s m^{-1}
second(s) per metre	s m^{-1}		$= 1$ s m^{-1}

9.20 Dimension LT^{-1}

9.20.1 $\langle 2 \rangle$ **length rate**
$\langle 3 \rangle$ **speed**
$\langle 5 \rangle$ v, c
$\langle 7 \rangle$ $v = $ dl/dt

9.20.1.1 $\langle 2 \rangle$ **length rate vector**
$\langle 3 \rangle$ **velocity**
$\langle 4 \rangle$ vel.
$\langle 5 \rangle$ $\boldsymbol{v}$, $\boldsymbol{c}$, (u,v,w)
$\langle 6 \rangle$ change in position divided by time of travel
$\langle 7 \rangle$ $\boldsymbol{v} = $ d$\boldsymbol{s}$/dt
$\qquad = $ d(x,y,z)/dt

9.20.1.2 $\langle 2 \rangle$ **length rate of migrating component** B in system 1
$\langle 3 \rangle$ **rate of migration; velocity of migration; rate of travel**
$\langle 5 \rangle$ v, v_B
$\langle 6 \rangle$ change in position of a component divided by time of travel
$\langle 7 \rangle$ $v_B = $ dl_B/dt

⟨9⟩ The term **migration** is used in electrophoresis when movement of an ionic component is due to an electrical field. In planar chromatography and similar techniques for separation, the term **travel** is used.

⟨10⟩ Sem—Spermatozoa; vel.(curvilinear) $= 25~\mu\mathrm{m}~\mathrm{s}^{-1}$

⟨19⟩ The kind-of-quantity 'velocity' is commonly termed "motility" in spermiology.

9.20.1.3 ⟨2⟩ **length rate vector of suspended component** B in fluid 1 in a force field
⟨3⟩ **sedimentation velocity**
⟨5⟩ $\boldsymbol{v}_\mathrm{B}$
⟨6⟩ change of position of a component with respect to the fluid 1 in the direction of gravitational or centrifugal acceleration divided by time of travel

9.20.1.4 ⟨2⟩ **length rate of electromagnetic radiation**
⟨3⟩ **speed of propagation of electromagnetic radiation**; phase velocity of electromagnetic radiation; phase speed of electromagnetic waves; speed of electromagnetic waves; velocity of electromagnetic waves
⟨5⟩ c
⟨8⟩ speed of propagation of waves; velocity of propagation of waves
⟨9⟩ (1) The speed of electromagnetic radiation in a vacuum can be distinguished with the subscript zero, c_0 (§9.20U).
⟨9⟩ (2) For a transmitted wave, speed of propagation is the product of wavelength, λ, and frequency, ν:

$$c = \lambda\,\nu$$

9.20.2   ⟨2⟩ **areic volume rate** of component B from system 1
⟨5⟩ $\varphi_{V,\mathrm{B}}, J_{V,\mathrm{B}}$
⟨6⟩ volume rate of a component divided by an area of the system, commonly the area transverse to the direction of flow
⟨7⟩ $\varphi_{V,\mathrm{B}} = \Phi_\mathrm{B}/A_1$
⟨8⟩ volume flux

9.20U   **Units of dimension LT^{-1}**
⟨12⟩ *Coherent SI unit* metre(s) per second (m s^{-1})

Term for unit	Symbol of unit Recommended	Other	Value in coherent SI unit
speed of electromagnetic radiation in a vacuum	c_0	c	$= 299.792\,458 \times 10^3$ m s^{-1}
kilometre(s) per second	km s^{-1}		$= 1000$ m s^{-1}
metre(s) per second	m s^{-1}		$= 1$ m s^{-1}
kilometre(s) per hour	km h^{-1}		$= (1/3.6)$ m s^{-1} $\approx 277.778 \times 10^{-3}$ m s^{-1}
metre(s) per minute		m min^{-1}	$\approx 16.666\,67 \times 10^{-3}$ m s^{-1}
centimetre(s) per second		cm s^{-1}	$= 0.01$ m s^{-1}
millimetre(s) per second	mm s^{-1}		$= 0.001$ m s^{-1}
metre(s) per hour		m h^{-1}	$\approx 0.277\,777\,8 \times 10^{-3}$ m s^{-1}
metre(s) per day	m d^{-1}		$\approx 11.574\,07 \times 10^{-6}$ m s^{-1}
micrometre(s) per second	μm s^{-1}		$= 10^{-6}$ m s^{-1}
metre(s) per year	m a^{-1}	m year^{-1}	$\approx 31.688\,76 \times 10^{-9}$ m s^{-1}
nanometre(s) per second	nm s^{-1}		$= 10^{-9}$ m s^{-1}

9.21 Dimension $L^{-2}T^{-1}$

9.21.1   ⟨2⟩ **areic absorbed dose** of component B in system 1
⟨5⟩ A_B/A_S
⟨7⟩ A_B/A_S,   where A_B is absorbed dose and A_S is area of a surface.
⟨8⟩ radioactive fall-out

9.21U   **Units of dimension $L^{-2}T^{-1}$**
⟨12⟩ *Coherent SI units* becquerel(s) per square metre or (one) per square metre second or reciprocal square metre second(s) (Bq m^{-2} = m^{-2} s^{-1})

9.22 Dimension $L^{-2}T$

9.22.1   ⟨2⟩ **areic time** of operation or process B in system 1
⟨5⟩ t_B/A_1

9.22.2   ⟨2⟩ **resistance to diffusion** of component B in system 1
⟨5⟩ D_B^{-1}
⟨6⟩ reciprocal of diffusion coefficient (§9.23.2)

9.22U   **Unit of dimension $L^{-2}T$**
⟨12⟩ *Coherent SI unit* second(s) per square metre (s m^{-2})

9.23 Dimension L^2T^{-1}

9.23.1   ⟨2⟩ **kinematic viscosity** of fluid 1
⟨5⟩ ν
⟨6⟩ dynamic viscosity of a fluid, η_1, divided by its volumic mass, ρ_1
⟨7⟩ $\nu_1 = \eta_1/\rho_1$
⟨10⟩ Pt—Plasma; kinematic viscosity(37 °C) = 1.25 mm^2 s^{-1}
⟨10⟩ Pt—Semen; kinematic viscosity(proc.) = 1.72 mm^2 s^{-1}

9.23.2   ⟨2⟩ **diffusion coefficient** of component B in system 1
⟨5⟩ D_B
⟨6⟩ absolute value of the product of local number concentration of a component, C_B, and local average velocity of molecules of that component, v_B, divided by the number concentration gradient ∇C_B
⟨7⟩ $D_B = |C_B \, v_B|/\nabla C_B$
⟨9⟩ Diffusion coefficients are scalar quantities.

9.23.3   ⟨2⟩ **thermal diffusivity**
⟨3⟩ **thermal diffusion coefficient**
⟨5⟩ a, D_T
⟨6⟩ (1) thermal conductivity of system 1, λ_1, divided by volumic mass of the system, ρ_1, and massic kelvic energy (or specific heat capacity) of the system at constant pressure, $c_{1,p}$
⟨7⟩ (1) $a_1 = \lambda_1/\rho_1 \, c_{1,p}$
⟨7⟩ (2) $D_T = k_T \, D$
⟨8⟩ D

9.23U   **Units of dimension L^{2}T^{-1}**

$\langle 12 \rangle$ *Coherent SI unit* square metre(s) per second (m^2 s^{-1})

Term for unit	Symbol of unit		Value in coherent SI unit
	Recommended	Other	
square metre(s) per second	m^2 s^{-1}		$= 1$ m^2 s^{-1}
square metre(s) per hour		m^2 h^{-1}	$= 277.777\,8 \times 10^{-6}$ m^2 s^{-1}
square centimetre(s) per second		cm^2 s^{-1}	$= 100 \times 10^{-6}$ m^2 s^{-1}
square millimetre(s) per second	mm^2 s^{-1}		$= 10^{-6}$ m^2 s^{-1}
square centimetre(s) per day		cm^2 d^{-1}	$= 1157.407 \times 10^{-12}$ m^2 s^{-1}

$\langle 19 \rangle$ Where the quantity is difficult to envisage as an area divided by a time, the unit is commonly termed, for instance, metre(s) squared per second, *e.g.* ISO/IEC 80000-4 [10] for kinematic viscosity, instead of square metre(s) per second.

9.24 Dimension L^{3}T^{-1}

9.24.1 $\langle 2 \rangle$ **volume rate** of component B in system 1
$\langle 3 \rangle$ volume flow rate; mean volume rate; volume rate of flow
$\langle 4 \rangle$ vol.rate
$\langle 5 \rangle$ $\dot{V}_B$, $\Phi_{V,B}$, $F_{V,B}$
$\langle 6 \rangle$ volume of component B changed in or moved to or from a system, ΔV_B, in a time, Δt, divided by that time
$\langle 7 \rangle$ (1) $\dot{V}_B = dV_B/dt$ in general
$\langle 7 \rangle$ (2) $\dot{V}_B = \Delta V_B/\Delta t$ for an average mass rate or a constant process
$\langle 8 \rangle$ volume flux; flux; flow; discharge; flow rate; discharge rate; volume per unit time; Φ_B, $q_{v,B}$, q_B, Q, U
$\langle 9 \rangle$ (1) The temperature and, for gases, the pressure of the isolated component should be stated. The value of the quantity depends on both conditions.
$\langle 9 \rangle$ (2) For volume rate, the system considered may be an interface across which matter is transferred, *e.g.* a cross-section or surface.
$\langle 10 \rangle$ Flame—Solution aspirated; vol.rate $= ?$ μL s^{-1}
$\langle 9 \rangle$ (3) aspiration flow; aspiration rate; Fv, qv
$\langle 10 \rangle$ Kidn—Water throughput; vol.rate(proc.) $= 10.5$ mL s^{-1}
$\langle 10 \rangle$ Kidn—Creatininium clearance; vol.rate(surf. $= 1.73$ m^2) $= 1.16$ mL s^{-1}
$\langle 10 \rangle$ Liver—Blood flow; vol.rate(proc.) $= 30$ mL s^{-1}

9.24.2 $\langle 2 \rangle$ **clearance** of component B from system 1 to output system 2
$\langle 5 \rangle$ C_B
$\langle 6 \rangle$ (1) product of the substance concentration of a component in a specified output fluid 2 and the volume rate of that output 2 divided by the substance concentration of B in system 1
$\langle 6 \rangle$ (2) volume of system 1 totally cleansed for component B in a time Δt divided by that range
$\langle 7 \rangle$ $C_B = c_{B,2}\, \dot{V}_2 / c_{B,1}$
$\langle 9 \rangle$ (1) In pulmonary toxicology, clearance is applied to the removal by ciliary action of an inhaled component that deposits on the lining surface of the lung expressed as volume of lungs effectively cleared divided by time.
$\langle 9 \rangle$ (2) In toxicology, clearance of component B from the organism may be calculated from the sum of the clearances through each eliminating organ or tissue for that component.

⟨9⟩ (3) This kind-of-quantity could, if possible, be replaced by the volume rate (*e.g.* of fluid filtered in the glomeruli).
⟨10⟩ Kidn—Lithium clearance; vol.rate(proc.) = 0.65 mL s^{-1}
⟨10⟩ Kidn—Chromium-EDTA clearance; vol.rate(proc.) = 1.78 mL s^{-1}

9.24U   Units of dimension $\mathbf{L^3 T^{-1}}$

⟨12⟩ *Coherent SI unit* cubic metre(s) per second (m^3 s^{-1})

Term for unit	Symbol of unit Recommended	Other	Value in coherent SI unit
cubic metre(s) per second	m^3 s^{-1}		$= 1$ m^3 s^{-1}
cubic metre(s) per minute		m^3 h^{-1}	$\approx 16.666\,67 \times 10^{-3}$ m^3 s^{-1}
cubic decimetre(s) per second		dm^3 s^{-1}	$= 10^{-3}$ m^3 s^{-1}
litre(s) per second	L s^{-1}, l s^{-1}		$= 10^{-3}$ m^3 s^{-1}
cubic metre(s) per hour		m^3 h^{-1}	$\approx 277.777\,8 \times 10^{-6}$ m^3 s^{-1}
litre(s) per minute		L min^{-1}, l min^{-1}	$\approx 16.666\,67 \times 10^{-6}$ m^3 s^{-1}
cubic metre(s) per day	m^3 d^{-1}		$\approx 11.574\,07 \times 10^{-6}$ m^3 s^{-1}
cubic centimetre(s) per second		cm^3 s^{-1}	$= 10^{-6}$ m^3 s^{-1}
millilitre(s) per second	mL s^{-1}, ml s^{-1}		$= 10^{-6}$ m^3 s^{-1}
litre(s) per hour		L h^{-1}, l h^{-1}	$\approx 277.777\,8 \times 10^{-9}$ m^3 s^{-1}
cubic metre(s) per year	m^3 a^{-1}	m^3 year^{-1}	$\approx 31.688\,76 \times 10^{-9}$ m^3 s^{-1}
litre(s) per day	L d^{-1}, l d^{-1}		$\approx 11.574\,07 \times 10^{-9}$ m^3 s^{-1}
cubic millimetre(s) per second	mm^3 s^{-1}		$= 10^{-9}$ m^3 s^{-1}
microlitre(s) per second	μL s^{-1}, μl s^{-1}		$= 10^{-9}$ m^3 s^{-1}

9.25   Dimension $\mathbf{M^{-1} T^{-1}}$

9.25.1   ⟨2⟩ **absorbed dose content** of component B in system 1
⟨5⟩ $a_{B,1}$
⟨6⟩ absorbed dose of a component, $A_{B,1}$, divided by the mass of the system, m_1
⟨7⟩ $a_{B,1} = A_{B,1}/m_1$
⟨8⟩ radioactivity content; specific activity
⟨9⟩ (1) For a pure radionuclide, A*, the property can be termed massic absorbed dose (§7.1, Table 7.2):

$$a_1 = a_{c^*,1} \quad \text{where the star indicates 'pure'.}$$

⟨9⟩ (2) The terms activity content and specific activity are ambiguous, being used also, for instance, of catalytic-activity content (§7.1, Table 7.1).

9.25U   Units of dimension $\mathbf{M^{-1} T^{-1}}$

⟨12⟩ *Coherent SI units* becquerel(s) per kilogram or reciprocal kilogram second(s) (Bq kg^{-1} = kg^{-1} s^{-1})

9.26   Dimension $\mathbf{M T^{-1}}$

9.26.1   ⟨2⟩ **mass rate** of component B in system 1
⟨3⟩ mean mass rate; mass flow rate; mass rate of flow; mass transfer rate; mass velocity
⟨5⟩ $\dot{m}_B$, $\Phi_{m,B}$, $F_{m,B}$, $J_{m,B}$

⟨6⟩ mass of a component changed in or moved to or from a system in a small time, $d_{m,B}$, divided by the difference in time, dt

⟨7⟩ (1) $\dot{m}_B = dm_B/dt$ in general

⟨7⟩ (2) $\langle \dot{m}_B \rangle = \Delta m_B/\Delta t$ for an average mass rate

⟨8⟩ mass flux; mass flow; flow rate; change of mass per unit time; $q_{m,B}$, G, Δ

⟨9⟩ (1) If volumic mass, ρ_1, is known, mass rate of the system by flow can be calculated from volume rate, $\dot{V}$:

$$\dot{m}_B = \dot{V}_1\, \rho_1$$

⟨9⟩ (2) For flow, the system may be defined as an interface or surface across which matter is transferred.

⟨9⟩ (3) Results can often be reported either as a mass rate or as a mass as long as times are specified (§6.8).

⟨9⟩ (4) The differential quotient and the difference quotient to time should be distinguished as rate and average rate, respectively. The distinction can also be indicated in the specifications of any quantity as the time or time difference of measurement (§6.8).

⟨9⟩ (5) Mass rate of a diffusing gas (*e.g.* water vapour) can be estimated by the equation:

$$\dot{m}_B = -D_\alpha\, A\, \nabla \rho_B$$

where D_α is mass diffusion coefficient of phase α, A is area of interface of diffusion and $\nabla \rho_B$ is mass concentration gradient.

⟨10⟩ Organism—Growth; mass rate(24 h) = 100 g d^{-1}

⟨10⟩ Pt—Protein(ingestion); mass rate(24 h) = 125 g d^{-1}

⟨10⟩ Pt(U)—Protein; mass rate(24 h) = 355 mg d^{-1}

⟨10⟩ Pt—Faeces; mass rate(24 h) = 0.350 kg d^{-1}

9.26.2   ⟨2⟩ **mechanical impedance** of component B at surface or point

⟨5⟩ $Z_{m,B}$

⟨6⟩ total force, F_x, divided by average velocity of entities of a component, $v_{x,B}$, at the surface or point in the direction of the force

⟨7⟩ $Z_{m,B} = F_x/v_{x,B}$

    $= p_1/\varphi_{V,B}$

⟨9⟩ (1) Although defined by ISO/IEC 80000-8,[11] for acoustics, the kind-of-quantity is also applicable, for instance, in pulmonary physiology for flow of respiratory gases.

⟨9⟩ (2) Mechanical impedance is a vector kind-of-quantity.

9.26U   **Units of dimension MT^{-1}**

⟨12⟩ *Coherent SI unit of mass rate* kilogram(s) per second (kg s^{-1})

⟨12⟩ *Coherent SI units of mechanical impedance* newton second(s) per metre or kilogram(s) per second (N s m^{-1} = kg s^{-1})

Term for unit	Symbol of unit		Value in coherent SI unit
	Recommended	Other	
kilogram(s) per second	kg s^{-1}		= 1 kg s^{-1}
megagram(s) per hour		Mg h^{-1}	≈ 277.777 8 × 10^{-3} kg s^{-1}
kilogram(s) per minute		kg min^{-1}	≈ 16.666 67 × 10^{-3} kg s^{-1}
megagram(s) per day	Mg d^{-1}		≈ 11.574 07 × 10^{-3} kg s^{-1}
gram(s) per second	g s^{-1}		= 10^{-3} kg s^{-1}
kilogram(s) per hour		kg h^{-1}	≈ 277.777 8 × 10^{-6} kg s^{-1}
gram(s) per minute		g min^{-1}	≈ 16.666 67 × 10^{-6} kg s^{-1}
kilogram(s) per day	kg d^{-1}		≈ 11.574 07 × 10^{-6} kg s^{-1}
milligram(s) per second	mg s^{-1}		= 10^{-6} kg s^{-1}
gram(s) per hour		g h^{-1}	≈ 277.777 8 × 10^{-9} kg s^{-1}
kilogram(s) per year	kg a^{-1}	kg year^{-1}	≈ 31.688 76 × 10^{-9} kg s^{-1}

(Continued)

Term for unit	Symbol of unit		Value in coherent SI unit
	Recommended	Other	
milligram(s) per minute		mg min^{-1}	$\approx 16.666\,67 \times 10^{-9}$ kg s^{-1}
gram(s) per day	g d^{-1}		$\approx 11.574\,07 \times 10^{-9}$ kg s^{-1}
microgram(s) per second	μg s^{-1}		$= 10^{-9}$ kg s^{-1}
milligram(s) per hour		mg h^{-1}	$\approx 277.777\,8 \times 10^{-12}$ kg s^{-1}
milligram(s) per day	mg d^{-1}		$\approx 11.574\,07 \times 10^{-12}$ g s^{-1}
nanogram(s) per second	ng s^{-1}		$= 10^{-12}$ kg s^{-1}

⟨19⟩ Because of the inconvenient conversion factors for comparison of results from different sources, the range of denominator units should be limited, for instance to second, day and year, although some other units are recognized for use with SI (§6.8 and §9.17U, (2)).

9.27 Dimension LMT^{-1}

9.27.1   ⟨2⟩ **momentum** of body 1
       ⟨5⟩ p, $\boldsymbol{p}$ (as vector), $\vec{p}$ (as vector)
       ⟨6⟩ product of the mass of a body, m_1, and its velocity, $\boldsymbol{v}_1$
       ⟨7⟩ $\boldsymbol{p} = m_1\,\boldsymbol{v}_1$

9.27.2   ⟨2⟩ **impulse** to body 1
       ⟨5⟩ I, $\boldsymbol{I}$ (as vector), $\vec{I}$ (as vector)
       ⟨7⟩ $\boldsymbol{I} = \int \boldsymbol{F}\,\mathrm{d}t$
       ⟨9⟩ For a time interval $[t_0/t_1]$

$$\boldsymbol{I} = \boldsymbol{p}(t_1) - \boldsymbol{p}(t_0)$$

9.27U   **Units of dimension LMT^{-1}**
       ⟨12⟩ *Coherent SI units of momentum* kilogram metre(s) per second or newton second(s) (kg m s^{-1} = N s)
       ⟨12⟩ *Coherent SI unit of impulse* newton second(s) (N s)

9.28 Dimension L^{-1}MT^{-1}

9.28.1   ⟨2⟩ **dynamic viscosity**
       ⟨3⟩ **coefficient of internal friction**
       ⟨5⟩ η
       ⟨6⟩ shear stress in the x,z plane in a fluid, $\tau_{x,z}$, divided by velocity gradient perpendicular to the plane of shear, ∇v_y
       ⟨7⟩ $\eta = \tau_{x,z}\nabla v_y$
       ⟨8⟩ viscosity; μ
       ⟨9⟩ The definition is valid for laminar flow in which $v_y = 0$
       ⟨10⟩ Pt—Plasma; dynamic viscosity(37 °C) = 3.5 mPa s^{-1}

9.28U   **Units of dimension L^{-1}MT^{-1}**
       ⟨12⟩ *Coherent SI units* pascal second(s) or newton second(s) per square metre or kilogram(s) per metre second (Pa s = N s m^{-2} = kg m^{-1} s^{-1})

Term for unit	Symbol of unit	Value in coherent SI unit
	Recommended	
kilopascal second(s)	kPa s	= 1000 Pa s
pascal second(s)	Pa s	= 1 Pa s
kilogram(s) per metre second	kg m^{-1} s^{-1}	= 1 Pa s
millipascal second(s)	mPa s	= 0.001 Pa s

9.29   Dimension $L^2M^{-1}T$

9.29.1 $\langle 2 \rangle$ **hydraulic permeability**
$\langle 5 \rangle$ L_p
$\langle 6 \rangle$ areic rate of a liquid component across an interface, $\varphi_{v,\mathrm{B}}$, divided by the pressure gradient of this liquid across the interface, $\nabla \rho$
$\langle 7 \rangle$ $L_\mathrm{p} = \varphi_{v,\mathrm{B}}/\nabla \rho$

9.29U   **Units of dimension $L^2M^{-1}T$**
$\langle 12 \rangle$ *Coherent SI units* metre(s) per pascal second or second square metre(s) per kilogram (m $\mathrm{Pa}^{-1}\ \mathrm{s}^{-1} = \mathrm{s}\ \mathrm{m}^2\ \mathrm{kg}^{-1}$)

Term for unit	Symbol of unit		Value in coherent SI unit
	Recommended	Other	
megagram(s) per square metre second	$\mathrm{Mg\ m^{-2}\ s^{-1}}$		$= 1000\ \mathrm{kg\ m^{-2}\ s^{-1}}$
gram(s) per square millimetre second		$\mathrm{g\,mm^{-2}\ s^{-1}}$	$= 1000\ \mathrm{kg\ m^{-2}\ s^{-1}}$
gram(s) per square centimetre second		$\mathrm{g\,cm^{-2}\ s^{-1}}$	$= 10\ \mathrm{kg\ m^{-2}\ s^{-1}}$
kilogram(s) per square metre second	$\mathrm{kg\ m^{-2}\ s^{-1}}$		$= 1\ \mathrm{kg\ m^{-2}\ s^{-1}}$
pascal second(s) per metre	$\mathrm{Pa\ s\ m^{-1}}$		$= 1\ \mathrm{Pa\ s\ m^{-1}}$
gram(s) per square centimetre hour		$\mathrm{g\,cm^{-2}\ h^{-1}}$	$\approx 2.777\,778 \times 10^{-3}\ \mathrm{kg\ m^{-2}\ s^{-1}}$
gram(s) per square metre second	$\mathrm{g\ m^{-2}\ s^{-1}}$		$= 1 \times 10^{-3}\ \mathrm{kg\ m^{-2}\ s^{-1}}$
gram(s) per square centimetre day		$\mathrm{g\,cm^{-2}\ d^{-1}}$	$\approx 115.740\,7 \times 10^{-6}\ \mathrm{kg\ m^{-2}\ s^{-1}}$
milligram(s) per square metre second	$\mathrm{mg\ m^{-2}\ s^{-1}}$		$= 10^{-5}\ \mathrm{kg\ m^{-2}\ s^{-1}}$
gram(s) per square metre day	$\mathrm{g\ m^{-2}\ d^{-1}}$		$\approx 11.574\,7 \times 10^{-9}\ \mathrm{kg\ m^{-2}\ s^{-1}}$
microgram(s) per square metre second	$\mathrm{\mu g\ m^{-2}\ s^{-1}}$		$= 10^{-9}\ \mathrm{kg\ m^{-2}\ s^{-1}}$
gram(s) per square metre year	$\mathrm{g\ m^{-2}\ a^{-1}}$	$\mathrm{g\ m^{-2}\,year^{-1}}$	$\approx 31.688\,76 \times 10^{-12}\ \mathrm{kg\ m^{-2}\ s^{-1}}$

9.30   Dimension $L^{-2}MT^{-1}$

9.30.1 $\langle 2 \rangle$ **areic mass rate** of component B in system 1
$\langle 5 \rangle$ $\varphi_{m,\mathrm{B}}$
$\langle 6 \rangle$ mass of a component transformed or transferred to or from the system during a time, $\mathrm{d}m_\mathrm{B}$, divided by that time, $\mathrm{d}t$, and by a specified area of the system, A
$\langle 7 \rangle$ $\varphi_{m,\mathrm{B}} = (\mathrm{d}m_\mathrm{B}/\mathrm{d}t)/A$
$\qquad = \dot{m}_\mathrm{B}/A$

9.30U   **Units of dimension $L^{-2}MT^{-1}$**
$\langle 12 \rangle$ *Coherent SI unit* kilogram(s) per square metre second $(\mathrm{kg\ m^{-2}\ s^{-1}})$

9.31   Dimension L^2MT^{-1}

9.31.1 $\langle 2 \rangle$ **angular momentum** of a body about a point
$\langle 3 \rangle$ **moment of momentum**

⟨5⟩ $L, \vec{L}, J$
⟨6⟩ vector product of the radius vector, r, from the point to the particle and the momentum of the particle, p
⟨7⟩ $L = r \times p$

9.31.2   ⟨2⟩ **Planck constant**
⟨5⟩ $h, \hbar$
⟨9⟩ (1) $h = 662.606\,957 \times 10^{-36}$ J s; $u_c = 0.000\,029 \times 10^{-36}$ J s. The assignment of a numerical value to the Planck constant is under discussion (CIPM, CCU), see §5.1.9.
⟨9⟩ (2) The symbol $\hbar$ is used for $h/(2\pi)$.

9.31U   **Units of dimension L^2MT^{-1}**
⟨12⟩ *Coherent SI unit of angular momentum* kilogram metre(s) squared per second $(\text{kg m}^2 \text{ s}^{-1} = \text{N m s})$
⟨12⟩ *Coherent SI of Planck constant* joule second(s) or joule(s) per hertz, or kilogram metre(s) squared per second $(\text{J s} = \text{J Hz}^{-1} = \text{kg m}^2 \text{ s}^{-1})$

9.32   Dimension $L^{-3}MT^{-1}$

9.32.1   ⟨2⟩ **mass concentration rate** of component B in system 1
⟨5⟩ $\dot{\rho}_{B,1}$
⟨6⟩ change in mass concentration of a component during a time, $d\rho_{B,1}$, divided by the time difference, dt
⟨7⟩ $\dot{\rho}_{B,1} = d\rho_{B,1}/dt$
⟨8⟩ rate of increase in biomass; production rate (biotechnology)
⟨9⟩ It is equal to the average mass concentration rate only if the system does not change in volume:

$$\dot{\rho}_{B,1} = \Delta\rho_{B,1}/\Delta t$$

$$= [\Delta(m_B/V_1)/\Delta t]_V$$

⟨10⟩ P—Urea; mass c.rate(proc.) $= 0.14$ mg L^{-1} s^{-1}

9.32U   **Units of dimension $L^{-3}MT^{-1}$**
⟨12⟩ *Coherent SI unit* kilogram(s) per cubic metre second $(\text{kg m}^{-3} \text{ s}^{-1})$

Term for unit	Symbol of unit Recommended	Other	Value in coherent SI unit
kilogram(s) per cubic metre second	kg m^{-3} s^{-1}		$= 1$ kg m^{-3} s^{-1}
gram(s) per litre second	g L^{-1} s^{-1}		$= 1$ kg m^{-3} s^{-1}
gram(s) per cubic metre second	g m^{-3} s^{-1}		$= 0.001$ kg m^{-3} s^{-1}
milligram(s) per litre second	mg L^{-1} s^{-1}		$= 0.001$ kg m^{-3} s^{-1}
milligram(s) per cubic metre second	mg m^{-3} s^{-1}		$= 10^{-6}$ kg m^{-3} s^{-1}
gram(s) per cubic metre hour		g m^{-3} h^{-1}	$\approx 277.777\,8 \times 10^{-9}$ kg m^{-3} s^{-1}

9.33 Dimension T^{-2}

9.33.1 $\langle 2 \rangle$ **angular acceleration**
$\langle 5 \rangle$ α
$\langle 6 \rangle$ change in angle rate, dω, during a small time, dt, divided by that time difference
$\langle 7 \rangle$ $\alpha = \mathrm{d}\omega/\mathrm{d}t$

9.33U   **Units of dimension T^{-2}**
$\langle 12 \rangle$ *Coherent SI units* radian(s) per second squared or (one) per second squared or reciprocal second(s) squared (rad s^{-2} = s^{-2})

9.34 Dimension LT^{-2}

9.34.1 $\langle 2 \rangle$ **linear acceleration** of body 1 or entity of component B
$\langle 3 \rangle$ **acceleration**
$\langle 5 \rangle$ a (as scalar), **a**, $\vec{a}$ (as vector)
$\langle 6 \rangle$ change in linear velocity, dv, during a small time, dt, divided by that time difference
$\langle 7 \rangle$ $\boldsymbol{a} = \mathrm{d}\boldsymbol{v}/\mathrm{d}t$ in general
$= \mathrm{d}^2(x,y,z)/\mathrm{d}t^2$ in general
$= \mathrm{d}^2s/(\mathrm{d}t)^2$ for rectilinear motion
$\langle 9 \rangle$ Linear acceleration is a vector quantity.
$\langle 10 \rangle$ Air—Sound particles; linear acceleration = ? m s^{-2}
$\langle 9 \rangle$ Synonyms: sound particle acceleration; instantaneous sound particle acceleration; a_a, a

9.34.1.1 $\langle 2 \rangle$ **centrifugal acceleration**
$\langle 5 \rangle$ $\boldsymbol{a}_\mathrm{rot}$
$\langle 6 \rangle$ acceleration of a body as a result of uniform rotational motion
$\langle 9 \rangle$ (1) Centrifugal acceleration is a vector kind-of-quantity.
$\langle 9 \rangle$ (2) For an example of calculation see §7.4.2.

9.34.1.2 $\langle 2 \rangle$ **acceleration due to gravity**
$\langle 3 \rangle$ **acceleration by gravity; acceleration of free fall**
$\langle 5 \rangle$ g (as scalar), **g**, $\vec{g}$ (as vector)
$\langle 6 \rangle$ acceleration of any body in free fall in a vacuum due to gravity
$\langle 9 \rangle$ Standard acceleration (due to gravity), g_n, assumes sea level and latitude 45°:

$$g_\mathrm{n} = 9.806\,65 \text{ m s}^{-2}$$

9.34.2 $\langle 2 \rangle$ **massic energy gradient** of fluid 1
$\langle 5 \rangle$ **grad** ψ, $\nabla\psi_m$
$\langle 6 \rangle$ change in massic energy of a fluid, ψ_m, in distance (x,y,z) divided by that distance
$\langle 7 \rangle$ $\nabla\psi_m = \mathrm{d}\psi_m/\mathrm{d}(x,y,z)$
$\langle 8 \rangle$ $\nabla\psi_M$; gradient of specific fluid potential
$\langle 10 \rangle$ Multiphase system—Water(l); massic energy gradient = ? m s^{-2}

9.34U   **Units of dimension LT^{-2}**
$\langle 12 \rangle$ *Coherent SI unit of acceleration* metre(s) per second squared (m s^{-2})
$\langle 12 \rangle$ *Coherent SI units of massic energy (or potential) gradient* joule(s) per kilogram metre or metre(s) per second squared (J kg^{-1} m^{-1} = m s^{-2})

Term for unit or constant	Symbol of unit or constant		Value in coherent SI unit
	Recommended	Other	
standard acceleration	g_n	g	= 9.806 65 m s^{-2}
metre(s) per second	m s^{-2}		= 1 m s^{-2}
newton(s) per kilogram	N kg^{-1}		= 1 m s^{-2}

(*Continued*)

Term for unit or constant	Symbol of unit or constant		Value in coherent SI unit
	Recommended	Other	
kilometre(s) per hour second		km h^{-1} s^{-1}	$\approx 0.277\,778$ m s^{-2}
centimetre(s) per second squared		cm s^{-2}	$= 0.01$ m s^{-2}
millimetre(s) per second squared	mm s^{-2}		$= 10^{-3}$ m s^{-2}

9.35 Dimension L^{-2}T^{-2}

9.35.1 $\langle 2 \rangle$ **areic absorbed dose rate** of component B in system 1
$\langle 3 \rangle$ **rate of fall-out**
$\langle 7 \rangle$ $(dA_B/dt)/A_1$, where A_B is absorbed dose of component B and A_1 is the area of the system.

9.35U   **Units of dimension L^{-2}T^{-2}**
$\langle 12 \rangle$ *Coherent SI units* becquerel(s) per square metre second or (one) per square metre second squared or reciprocal square metre(s) second squared (Bq m^{-2} s^{-1} = m^{-2} s^{-2})

9.36 Dimension L^{-2}T^{2}

9.36.1 $\langle 2 \rangle$ **mass–energy quotient** of system 1
$\langle 6 \rangle$ mass of a system divided by its energy
$\langle 9 \rangle$ The quotient widely used for fuels or foodstuffs is mass–enthalpy quotient of combustion.

9.36U   **Units of dimension L^{-2}T^{2}**
$\langle 12 \rangle$ *Coherent SI units* kilogram(s) per joule or second(s) squared per square metre (kg J^{-1} = s^2 m^{-2})

9.37 Dimension L^{2}T^{-2}

9.37.1 $\langle 2 \rangle$ **massic energy** of system 1
$\langle 5 \rangle$ e_1
$\langle 6 \rangle$ energy of a system, Q_1, divided by its mass, m_1
$\langle 7 \rangle$ $e_1 = Q_1/m_1$
$\langle 8 \rangle$ specific energy

9.37.1.1 $\langle 2 \rangle$ **massic thermodynamic energy** of system 1
$\langle 3 \rangle$ **massic internal energy; specific thermodynamic energy**
$\langle 5 \rangle$ u_1
$\langle 6 \rangle$ internal energy of a system, U_1, divided by its mass, m_1
$\langle 7 \rangle$ $u_1 = U_1/m_1$

9.37.1.2 $\langle 2 \rangle$ **massic enthalpy** of system 1
$\langle 5 \rangle$ h_1
$\langle 6 \rangle$ enthalpy of a system, H_1, divided by its mass, m_1
$\langle 7 \rangle$ $h_1 = H_1/m_1$
$\langle 8 \rangle$ specific enthalpy

9.37.1.3 $\langle 2 \rangle$ **massic Helmholtz energy** of system 1
$\langle 3 \rangle$ **massic Helmholtz free energy; massic Helmholtz function**; specific Helmholtz free energy; specific Helmholtz function
$\langle 5 \rangle$ a_1, f_1

⟨6⟩ Helmholtz energy of a system, A_1, divided by its mass, m_1
⟨7⟩ $a_1 = A_1/m_1$

9.37.1.4 ⟨2⟩ **massic Gibbs energy** of system 1
⟨3⟩ **massic Gibbs free energy**; specific Gibbs free energy; specific Gibbs function
⟨5⟩ g_1
⟨6⟩ Gibbs energy of a system, G_1, divided by its mass, m_1
⟨7⟩ $g_1 = G_1/m_1$

9.37.1.5 ⟨2⟩ **massic energy of ionizing radiation λ absorbed** into system 1
⟨3⟩ massic absorbed dose of radiation
⟨5⟩ D
⟨6⟩ energy of the radiation absorbed by a system, $Q_{e,abs,1}$, commonly ionizing radiation, divided by the mass of the system, m_1
⟨7⟩ $D_1 = Q_{e,abs,1}/m_1$
⟨8⟩ absorbed dose of radiation; dose

9.37.1.6 ⟨2⟩ **effective massic energy equivalent of ionizing radiation λ** into system 1
⟨5⟩ H
⟨6⟩ massic energy of ionizing radiation absorbed by a system, D_1, multiplied by factors $w_{R,\lambda}$, and N_1 expressing the biological effects of exposure to different types of radiation and distributions of massic energy on a common scale
⟨7⟩ $H_1 = w_{R,\lambda} N_1 D_1$
⟨8⟩ dose equivalent
⟨9⟩ The factors w_R and N are defined in ref. 12.

9.37U **Units of dimension $L^2 T^{-2}$**
⟨12⟩ *Coherent SI units of massic energy* joule(s) per kilogram or square metre(s) per second squared ($J\ kg^{-1} = m^2\ s^{-2}$)
⟨12⟩ *Coherent SI units of massic energy of ionizing radiation* gray(s) or joule(s) per kilogram or square metre(s) per second squared ($Gy = J\ kg^{-1} = m^2\ s^{-2}$)
⟨12⟩ *Coherent SI units of effective massic energy of ionizing radiation* sievert(s) or joule(s) per kilogram or square metre(s) per second squared ($Sv = J\ kg^{-1} = m^2\ s^{-2}$)

Term for unit	Symbol of unit		Value in coherent SI unit
	Recommended	Other	
calorie(s) per gram		cal g^{-1}	$= 4186.8\ J\ kg^{-1}$
kilocalorie(s) per kilogram		kcal kg^{-1}	$= 4186.8\ J\ kg^{-1}$
kilocalorie(s) per kilogram (nutrition)		kcal$_{th}$ kg^{-1}	$= 4184\ J\ kg^{-1}$
large calorie(s) per kilogram		Cal kg^{-1}	$= 4184\ J\ kg^{-1}$
kilojoule(s) per kilogram	kJ kg^{-1}		$= 1000\ J\ kg^{-1}$
square metre(s) per second squared			$1\ m^2\ s^{-2}$
joule(s) per kilogram	J kg^{-1}		$= 1\ J\ kg^{-1}$
gray(s)	Gy		$= 1\ J\ kg^{-1}$
sievert(s)	Sv		$= 1\ J\ kg^{-1}$
rad		rad, rd	$= 0.01\ Gy$
rem		rem	$= 0.01\ Sv$
erg(s) per gram		erg g^{-1}	$= 0.1 \times 10^{-3}\ J\ kg^{-1}$

⟨19⟩ (1) The gray is a special unit of SI permitted by CIPM 'to safeguard human health'.
⟨19⟩ (2) rad is the unit of radiation applied dose and rem is the unit of radiation equivalent man.

9.38   Dimension MT^{-2}

9.38.1 $\langle 2 \rangle$ **areic energy**
$\langle 3 \rangle$ **energy fluence**
$\langle 5 \rangle$ ψ
$\langle 7 \rangle$ $\psi_1 = Q_1/A_1$ where Q_1 is a kind-of-quantity of system 1 and A_1 is the area of the
system 1.
$\langle 8 \rangle$ energy density

9.38.1.1 $\langle 2 \rangle$ **areic energy of interface** B
$\langle 3 \rangle$ **surface tension**; surface energy; interfacial energy
$\langle 5 \rangle$ γ, σ
$\langle 6 \rangle$ force perpendicular to a line element in a surface divided by the length of that
line element

9.38U   **Units of dimension MT^{-2}**
$\langle 12 \rangle$ *Coherent SI units of areic energy and of energy of interface* joule(s) per square metre
or newton(s) per metre or kilogram(s) per second squared $(\text{J m}^{-2} = \text{N m}^{-1} = \text{kg s}^{-2})$

Term for unit	Symbol of unit Recommended	Other	Value in coherent SI unit
kilogram(s)-force per centimetre		kgf cm^{-1}	$= 980.665 \text{ N m}^{-1}$
joule(s) per square metre	J m^{-2}		$= 1 \text{ J m}^{-2}$
newton(s) per metre	N m^{-1}		$= 1 \text{ N m}^{-1}$
kilogram(s) per second squared	kg s^{-2}		$= 1 \text{ kg s}^{-2}$

9.39   Dimension LMT^{-2}

9.39.1 $\langle 2 \rangle$ **force acting** on body 1
$\langle 5 \rangle$ F_1 (as scalar), F, $\vec{F}$ (as vector)
$\langle 6 \rangle$ (1) change in momentum of a body, $\boldsymbol{m}_1$, in a small time, t, divided by the time
difference
$\langle 7 \rangle$ $F_1 = \mathrm{d}\boldsymbol{m}_1/\mathrm{d}t$
$\langle 6 \rangle$ (2) product of the mass of a body, m_1, and its acceleration, $\boldsymbol{a}_1$
$\langle 7 \rangle$ $F_1 = m_1 \, \boldsymbol{a}_1$
$= m_1 \, \mathrm{d}\boldsymbol{v}_1/\mathrm{d}t$

9.39.1.1 $\langle 2 \rangle$ **force due to gravity** on body 1 in a specified reference system
$\langle 5 \rangle$ F_g (as scalar), $\boldsymbol{F}_g$, $\vec{F}_g$ (as vector)
$\langle 6 \rangle$ force that would give a body an acceleration equal to local acceleration of free
fall in a specified reference system
$\langle 8 \rangle$ weight; G, P, W
$\langle 9 \rangle$ 'Weight' in this sense depends on local gravitational force and local centrifugal
force due to rotation of the earth. It may be measured with a spring balance.
However, in common parlance 'weight' often means mass (§9.8.1).
$\langle 10 \rangle$ Person $(m = 70 \text{ kg})$—Body; force $(g_n) = 686.465\,5 \text{ N}$

9.39.1.2 $\langle 2 \rangle$ **centrifugal force** on body 1
$\langle 5 \rangle$ F_{rot} (as scalar), $\boldsymbol{F}_{\text{rot}}$ (as vector)
$\langle 6 \rangle$ (1) force acting on a body as a result of centrifugal acceleration
$\langle 6 \rangle$ (2) product of the mass of a body, m_1, and its centrifugal acceleration, $\boldsymbol{a}_{\text{rot},1}$
$\langle 7 \rangle$ $\boldsymbol{F}_{\text{rot},1} = m_1 \, \boldsymbol{a}_{\text{rot},1}$

9.39.2 $\langle 2 \rangle$ **wavelength differential function of energy** of radiation λ
           $\langle 5 \rangle$ $Q_{e,\lambda}$
           $\langle 7 \rangle$ $Q_{e,\lambda} = dQ_e(\delta\lambda)/d\lambda$
           $\langle 8 \rangle$ spectral radiant energy
           $\langle 9 \rangle$ See §5.6.4.

9.39U   **Units of dimension LMT^{-2}**
          $\langle 12 \rangle$ *Coherent SI units* newton(s) or joule(s) per metre or kilogram metre(s) per second squared $(\mathrm{N} = \mathrm{J}\,\mathrm{m}^{-1} = \mathrm{kg}\,\mathrm{m}\,\mathrm{s}^{-2})$

	Symbol of unit		
Term for unit	Recommended	Other	Value in coherent SI unit
kilonewton(s)	kN		$= 1000\ \mathrm{N}$
calorie(s) per centimetre		$\mathrm{cal\,cm}^{-1}$	$= 418.68\ \mathrm{J\,m}^{-1}$
kilogram(s)-force		kgf	$= 9.806\,65\ \mathrm{N}$
newton(s)	N		$= 1\ \mathrm{N}$
joule(s) per metre	$\mathrm{J\,m}^{-1}$		$= 1\ \mathrm{J\,m}^{-1}$
kilogram metre(s) per second squared	$\mathrm{kg\,m\,s}^{-2}$		$= 1\ \mathrm{kg\,m\,s}^{-2}$
centimetre gram(s) per second squared		$\mathrm{cm\,g\,s}^{-2}$	$= 10^{-5}\ \mathrm{N}$
electronvolt(s) per centimetre		$\mathrm{eV\,cm}^{-1}$	$\approx 16.021\,765\,65 \times 10^{-18}\ \mathrm{J\,m}^{-1}$ $u_c = 0.000\,000\,35 \times 10^{-18}\ \mathrm{J\,m}^{-1}$

9.40   Dimension LM^{-1}T^2

9.40.1 $\langle 2 \rangle$ **substance-fractional solubility coefficient** of dissolving gas B in solution 1
           $\langle 3 \rangle$ **rational solubility coefficient**
           $\langle 5 \rangle$ $\alpha_{x,\mathrm{B}}$
           $\langle 6 \rangle$ substance fraction of a component B in a solution, x_B, divided by the active pressure in a gas phase at equilibrium, $\tilde{p}_\mathrm{B}$
           $\langle 7 \rangle$ $\alpha_{x,\mathrm{B}} = x_\mathrm{B}/\tilde{p}_\mathrm{B}$
                    $= \alpha_{x,\mathrm{B},\infty}/f_{x,\mathrm{B}}$ where $f_{x,\mathrm{B}}$ is activity coefficient (§8.10.3.1).
           $\langle 9 \rangle$ (1) $\alpha_{x,\mathrm{B},\infty} = \alpha_{x,\mathrm{B}}/\tilde{p}\mathrm{B}$
           $\langle 9 \rangle$ (2) For solvent A,
                    $\alpha_{x,\mathrm{A}} = a_{x,\mathrm{A}*}/f_\mathrm{A}$ where $\alpha_{x,\mathrm{A}*} = 1/\tilde{p}_\mathrm{A}$ and where the star indicates 'pure'.

9.40U   **Units of dimension LM^{-1}T^2**
          $\langle 12 \rangle$ *Coherent SI units* (one) per pascal or reciprocal pascal(s) or metre(s) second squared per kilogram $(\mathrm{Pa}^{-1} = \mathrm{m}\,\mathrm{s}^2\,\mathrm{kg}^{-1})$

9.41   Dimension L^{-1}MT^{-2}

9.41.1 $\langle 2 \rangle$ **pressure** of fluid system 1
           $\langle 3 \rangle$ **absolute pressure; total pressure** (of gas mixture); volumic hydraulic energy; areic force
           $\langle 4 \rangle$ pr.
           $\langle 5 \rangle$ p, p_t
           $\langle 6 \rangle$ (1) force, as a scalar quantity, $F_\perp$, exerted by a fluid normal to a plane divided by the area of that plane, A
           $\langle 7 \rangle$ $p = F_\perp/A$
           $\langle 6 \rangle$ (2) (phys. chem.) negative value of the partial differential of the internal energy of a system, U_1, divided by its volume, V_1

⟨7⟩ $p_1 = -(dU_1/dV_1)_{S(1),Q(1),n(B), ...}$

⟨8⟩ hydrostatic volumic energy; hydraulic pressure; hydrostatic pressure; P

⟨9⟩ (1) *Pressure* without qualification should be used for **total pressure** and not for **overpressure** (§9.41.1.2).

⟨9⟩ (2) An actually measured height of a liquid requires correction for the temperature of the liquid and the local acceleration of free fall before conversion of a hydraulic height to a pressure (§9.41U, ⟨19⟩ (1)).

⟨10⟩ Alveoli—Gas; pr. $= 101.3$ kPa

⟨10⟩ Artery—Blood; pr.(systolic, diastolic) $= (16.4, 10.8)$ kPa

⟨9⟩ Synonyms: **arterial pressure**; **blood pressure** are customary synonyms; p_{bl}, BP, b.p.

⟨10⟩ Room—Air; pr. $= 110$ kPa

⟨9⟩ Synonyms: **atmospheric** (or **air**) **pressure**; **ambient pressure** are customary synonyms; p_{amb}, p_a

9.41.1.1 ⟨2⟩ static pressure

⟨5⟩ p_s

⟨6⟩ pressure that would exist in the absence of fluid motion

9.41.1.2 ⟨2⟩ pressure difference

⟨3⟩ **excess pressure; overpressure**

⟨5⟩ Δp

⟨8⟩ ΔP, p_e, $p_{\text{Ü}}$

⟨9⟩ (1) Underpressure is a negative pressure difference; $-\Delta p$

⟨9⟩ (2) The symbol p_e is used for gauge pressure (or overpressure):

$$p_e = p - p_{amb} \quad \text{where } p_{amb} \text{ is the ambient pressure.}$$

Thus the gauge pressure is positive or negative depending on whether p is larger or smaller than p_{amb}, respectively.

9.41.1.3 ⟨2⟩ partial pressure of component B in gaseous system 1

⟨4⟩ part.pr.

⟨5⟩ p_B, $p(B)$

⟨6⟩ product of the substance fraction of a component, x_B, and the pressure of the gaseous system, p_1

⟨7⟩ $p_B = x_B\, p_1$

⟨8⟩ p, P

⟨9⟩ Elementary entities should be specified whenever doubt may arise.

⟨10⟩ Alveolar gas—Carbon dioxide; part.pr. $= 5.4$ kPa

⟨9⟩ Synonym: partial pressure of CO_2; $p(CO_2)$

⟨10⟩ Air—Dioxygen; part.pr.; $p(O_2)$

⟨10⟩ Air—Component(gas); part.pr. **vapour pressure**; p_B

⟨10⟩ Air—Water(gas); part.pr $= ?$ kPa

⟨9⟩ Synonyms: **water vapour pressure**; **vapour pressure**; p_w; p_v; e

⟨10⟩ Interface(liquid/gas)—Water(gas); part.pr.; $p(H_2O,i)$; e_i

⟨10⟩ Air(saturated)—Water vapour; part.pr.($20\ °C$) $= ?$ kPa

⟨9⟩ Synonyms: **water vapour pressure at saturation**; saturation vapour pressure; e_{sat}; e_s

⟨10⟩ Air—Water vapour; part.pr. deficit $= ?$ kPa

⟨9⟩ Synonym: **vapour pressure deficit**; Δe

9.41.2 ⟨2⟩ tension of gas B in liquid 1

⟨3⟩ **gas tension**

⟨5⟩ $p_{B,l}$, p_B

⟨6⟩ partial pressure of a component in a gas phase in equilibrium with the same component dissolved in a liquid phase
⟨8⟩ vapour tension
⟨9⟩ (1) Although gas tension is much used by clinical chemists and physiologists, substance concentration might be a more suitable kind-of-quantity for specification of dissolved gases.
⟨9⟩ (2) The decadic logarithm of vapour pressure is also termed *vapour tension*; pF.
⟨10⟩ P(arterial Blood)—Carbon dioxide; gas tension $= 5.65$ kPa

9.41.3   ⟨2⟩ **active pressure** of component B in gaseous system 1
⟨3⟩ **fugacity**
⟨5⟩ $\tilde{p}_B, f_B$
⟨7⟩ $\tilde{p}_B = g_B\, p_B$ where g_B is activity factor (§8.10.3).
⟨9⟩ (1) Active pressure is proportional to the chemical activity, λ_B, the factor of proportionality, g_B, being a function only of temperature and tending towards 1 at infinite dilution of the component
$$\tilde{p}_B = \lambda_B \lim_{x(B)\to 0} (x_B\, \tilde{p}/\lambda_B)$$
⟨9⟩ (2) $\lim_{p\to 0} \tilde{p}_B = p_B$

9.41.4   ⟨2⟩ **volumic energy** of system 1
⟨3⟩ **energy density**
⟨5⟩ w
⟨6⟩ mean energy in a system, Q_1, for instance from a radiation or a sound wave, divided by the volume of the system, V_1
⟨7⟩ $w_1 = Q_1/V_1$
⟨8⟩ density

9.41.4.1 ⟨2⟩ **osmotic volumic energy** of solution 1
⟨3⟩ **osmotic pressure**
⟨5⟩ Π, ψ
⟨6⟩ (1) pressure difference between a solution and the pure solvent A that provides the same chemical potential of the solvent in the solution and in the pure solvent
⟨6⟩ (2) excess pressure required to maintain osmotic equilibrium between the solution and the pure solvent A* separated by a membrane permeable only to the solvent
⟨7⟩ $\Pi = \int_{\mu(A^*)}^{\mu(A)} V_{m,A}^{-1}\,d\mu A$ where the star indicates 'pure'.
⟨8⟩ fluid potential
⟨9⟩ Integration assuming constant molar volume of solute:
$$V_{m,A} = V_{m,A^*}\quad \text{where the star indicates 'pure'.}$$
gives the van't Hoff equation:
$$\Pi = -(\mu_A - \mu_{A^*})/V_{A^*}$$
$$= R\,T\,\hat{c}$$

9.41.5   ⟨2⟩ **normal stress**
⟨5⟩ σ
⟨9⟩ In materials mechanics, stress is equated with pressure.

9.41.6   ⟨2⟩ **shear stress**
⟨5⟩ τ

9.41.7   ⟨2⟩ **cubic pressure coefficient** of system 1
⟨5⟩ K_1
⟨6⟩ small change in pressure of a gas divided by the resulting relative change in volume, other conditions being constant
⟨7⟩ $K_1 = -(\partial p/\partial V)\, V_1$
⟨8⟩ pressure coefficient

9.41.8   ⟨2⟩ **equilibrium coefficient** (based on partial pressure) of chemical reaction B
⟨3⟩ **baric equilibrium product**
⟨5⟩ K_p
⟨8⟩ equilibrium constant
⟨9⟩ The coefficient has the dimension $(\mathsf{L}^{-1}\mathsf{MT}^{-2})^n$, depending on the type of reaction.

9.41U   **Units of dimension $\mathsf{L}^{-1}\mathsf{MT}^{-2}$**
⟨12⟩ *Coherent SI units* pascal(s) or joule(s) per cubic metre or newton(s) per square metre or kilogram(s) per metre second squared (Pa = J m^{-3} = N m^{-2} = kg m^{-1} s^{-2})

Term for unit	Symbol of unit Recommended	Other	Value in coherent SI unit
megapascal(s)	MPa		$= 10^6$ Pa
standard atmosphere		atm	$= 101\,325$ Pa
normal pressure		p_n, p_0	$= 101\,325$ Pa
bar(s)		bar	$= 100 \times 10^3$ Pa
kilopascal(s)	kPa		$= 1000$ Pa
millimetre(s) head of mercury		mmHg	$\approx 133.322\,39$ Pa
torr(s)		Torr	$= (1/760)$ atm
			$\approx 133.322\,39$ Pa
millibar(s)		mbar, mb	$= 100$ Pa
kilogram(s)-force per square metre		kgf m^{-2}	$= 9.806\,65$ Pa
pascal(s)	Pa		$= 1$ Pa
newton(s) per square metre	N m^{-2}		$= 1$ Pa
joule(s) per cubic metre	J m^{-3}		$= 1$ Pa
kilogram(s) per metre second squared	kg m^{-1} s^{-2}		$= 1$ kg m^{-1} s^{-2}

⟨19⟩ (1) The pressure of a column of volumic mass ρ, of height h in a gravitational field of acceleration g is governed by the relationship:

$$p = g\, \rho\, h$$

The millimetre head of mercury is defined in terms of gravity, $g_n = 9.806\,65$ m s^{-2}. Volumic mass of mercury, $\rho(\text{Hg})$, is taken as $13.595\,1$ Mg m^{-3}, and of water, $\rho(\text{H}_2\text{O})$, as 1 Mg m^{-3}. For conversion of a fluid head, corrections are needed for local acceleration of gravity and for volumic mass at the temperature of measurement. Fluid head is not suitable for accurate measurement of pressure.

⟨19⟩ (2) To meet the wishes of cardiologists, the 30th and 34th World Health Assemblies of 1977 and 1981 recommended that the millimetre of mercury be retained alongside the kilopascal for expression of blood pressure, in the words of the 34th WHA, "until a future World Health Assembly considers the retention of the millimetre of mercury unnecessary for the undisturbed delivery of health care and the interchange of scientific information". The millimetre of mercury has not, however, been included by CGPM as a unit for use alongside SI since the 8th edition of the SI guide[2] and is therefore legally obsolete, except for some countries.

⟨19⟩ (3) The symbol b for bar is now deprecated because it conflicts with the use of b as a unit symbol to denote the barn, a unit of area.

9.42 Dimension L²MT⁻²

9.42.1 ⟨2⟩ **energy**
⟨3⟩ **amount-of-energy**
⟨5⟩ Q, E
⟨9⟩ The Einstein equation:

$$E = m\, c_0{}^2$$

is the fundamental relationship between the base kind-of-quantity mass, m, and energy, where c_0 is the speed of light in vacuum. With the Einstein equation, the special forms kinetic energy and potential energy can be defined.

9.42.1.1 ⟨2⟩ **kinetic energy** of body 1 in uniform motion
⟨5⟩ E_k, E_{kin}, T
⟨6⟩ (1) half the product of the mass of a body and the square of its velocity
⟨7⟩ $E_{k,1} = 0.5\, m_1\, v_1{}^2$
⟨6⟩ (2) energy due to uniform motion
⟨6⟩ (3) integral product of the velocity of a body, v_1, and its momentum, p_1
⟨7⟩ $dE_{k,1} = \int v_1\, dp_1$
⟨9⟩ For an example of calculation of kinetic energy see §7.4.3.

9.42.1.2 ⟨2⟩ **work**
⟨5⟩ W
⟨6⟩ integral of force acting on a body (§9.39.1) and distance travelled by the body (9.1.1.6)
⟨7⟩ $W = \int F\, dr$
⟨8⟩ A

9.42.1.3 ⟨2⟩ **potential energy**
⟨5⟩ E_{pot}
⟨7⟩ (1) $dE_{pot} = F\,(\partial x, \partial y, \partial z)$ where F is a force.
⟨7⟩ (2) $E_{pot} = \int F\, dr$ where F is a conservative force.

9.42.1.4 ⟨2⟩ **Gibbs energy**
⟨3⟩ **Gibbs free energy; Gibbs function**
⟨5⟩ G
⟨7⟩ $G = U + p \times V - T \times S$
$\quad = H - T \times S$
⟨9⟩ $dG = U + dp \times V - dT \times S + \varphi\, dQ + \Sigma(\mu_B\, dn_B)$

9.42.1.5 ⟨2⟩ **thermodynamic energy** of system 1 with process B
⟨3⟩ **internal energy**
⟨5⟩ U_1
⟨6⟩ (1) sum of amount-of-heat transferred to a system, ΔQ_1, and the work done on the system, W_1
⟨7⟩ (1) $\Delta U_1 = \Delta Q_1 + W_1$
⟨7⟩ (2) $dU_1 = -(p_1\, dV_1) + T_1\, dS_1 + \varphi_{1,2}\, d\rho + \Sigma(\mu_B\, dn_B)$

9.42.1.6 ⟨2⟩ **Helmholtz energy** of system 1
⟨3⟩ **Helmholtz free energy; Helmholtz function**
⟨5⟩ A_1, F_1
⟨7⟩ $A_1 = U_1 - T_1 \times S_1$

9.42.1.7 ⟨2⟩ **energy of radiation** $\Delta\lambda$
⟨3⟩ **radiant energy**

⟨5⟩ $Q_e(\Delta\lambda)$, Q
⟨6⟩ energy propagated as electromagnetic radiation

9.42.1.8 ⟨2⟩ **enthalpy** of system 1
⟨5⟩ H_1
⟨6⟩ sum of the thermodynamic (or internal) energy of a system, U_1, and the product of the pressure, p_1, and volume, V_1, of the system
⟨7⟩ $H_1 = U_1 + p_1 \times V_1$

9.42.1.9 ⟨2⟩ **amount-of-heat** in system 1
⟨3⟩ **quantity of heat**
⟨5⟩ Q_1
⟨8⟩ heat
⟨9⟩ Isothermal phase transformations, formerly termed *latent heat, L*, should be described in terms of thermodynamic functions such as the product of thermodynamic temperature and change in entropy, $T \times \Delta S$, or enthalpy, H.

9.42.2 ⟨2⟩ **moment of force** about a point
⟨5⟩ M (as scalar), $\boldsymbol{M}$, $\vec{M}$ (as vector)
⟨6⟩ vector product of any radius vector from a point to another point on the line of action of a force, $\boldsymbol{r}$, multiplied vectorially by the force, $\boldsymbol{F}$
⟨7⟩ $\boldsymbol{M} = \boldsymbol{r} \times \boldsymbol{F}$

9.42.3 ⟨2⟩ **differential quotient of energy rate to frequency** of radiation λ
⟨3⟩ spectral concentration of radiant power in terms of frequency; spectral radiant power in terms of frequency
⟨5⟩ P_ν, $\Phi_{e,\nu}$, Φ_ν
⟨8⟩ spectral concentration of radiant power (or flux); spectral density of radiant power (or flux)
⟨9⟩ Definition and practical usage are analogous to the differential quotient of energy rate to wavelength (§9.47.1).

9.42.4 ⟨2⟩ **entitic energy of photons** of radiation λ
⟨5⟩ $h\nu$
⟨6⟩ energy of a radiation divided by its number of photons
⟨7⟩ $h\nu = Q_e/N_\Phi$
⟨8⟩ photon energy; quantum energy
⟨9⟩ (1) The terms photon energy and quantum energy are also used for molar energy of photons (§9.104.1.6).
⟨9⟩ (2) Entitic energy of photons is independent of the medium of a radiation.
⟨9⟩ (3) For its use in specification of a given radiation see §7.3.12 and Table 7.3.

9.42U **Units of dimension L²MT⁻²**
⟨12⟩ *Coherent SI units of energy* joule(s) or kilogram square metre(s) per second squared (J = kg m² s⁻²)
⟨12⟩ *Coherent SI units of moment of force* newton metre(s) or joule(s), or kilogram square metre(s) per second squared (N m = J = kg m² s⁻²)
⟨12⟩ *Coherent SI units of differential quotient of energy rate to frequency* watt(s) per hertz or watt second (s) or joule(s), or kilogram square metre(s) per second squared (W Hz⁻¹ = W s = J = kg m² s⁻²)

⟨12⟩ *Coherent SI units of entitic energy of photons* watt(s) per number of photons (W)

Term for unit	Symbol of unit		Value in coherent SI unit
	Recommended	Other	
petajoule(s)	PJ		$= 10^{15}$ J
terajoule(s)	TJ		$= 10^{12}$ J
gigajoule(s)	GJ		$= 10^9$ J
kilowatt hour(s)		kW h	$= 3.6 \times 10^6$ J
megajoule(s)	MJ		$= 10^6$ J
kilocalorie(s)		kcal	$= 4186.8$ J
frigorie(s)		fg	$= 4186.8$ J
kilogram-calorie(s)		kg-cal	$= 4186.8$ J
Calorie(s) (nutrition)		kcal, Cal	≈ 4184 J
kilojoule(s)	kJ		$= 1000$ J
litre (1901) atmosphere(s)		l atm	$\approx 101.327\,8$ J
kilogram-force metre(s)		kgf m, kg m	$= 9.806\,65$ N m
international calorie(s)		cal, $\mathrm{cal_{IT}}$	$= 4.186\,8$ J
thermochemical calorie(s)		cal, $\mathrm{cal_{th}}$	$= 4.184$ J
erg(s)		erg	$= 10^{-7}$ J
joule(s)	J		$= 1$ J
watt(s) per hertz	W $\mathrm{Hz^{-1}}$		$= 1$ J
watt second(s)	W s		$= 1$ W $\mathrm{Hz^{-1}} = 1$ J
newton metre(s)	N m		$= 1$ N m $= 1$ J
millijoule(s)	mJ		$= 0.001$ J
microjoule(s)	μJ		$= 10^{-6}$ J
nanojoule(s)	nJ		$= 10^{-9}$ J
picojoule(s)	pJ		$= 10^{-12}$ J
femtojoule(s)	fJ		$= 10^{-15}$ J
attojoule (s)	aJ		$= 10^{-18}$ J
zeptojoule(s)	zJ		$= 10^{-21}$ J
electronvolt(s)	eV		$\approx 160.217\,656\,5 \times 10^{-21}$ J
			$u_c = 0.000\,003\,5 \times 10^{-21}$ J

⟨19⟩ The electronvolt is the kinetic energy acquired by an electron passing through an electric potential difference of 1 V in vacuum.

9.43   Dimension $\mathbf{L^3 M^{-1} T^{-2}}$

9.43.1   ⟨2⟩ **gravitational constant**
⟨5⟩ G
⟨6⟩ product of the masses of any two bodies, m_1 and m_2, divided by their distance apart, $l_{1,2}$, and by the gravitational force between them, $F_{1,2}$
⟨7⟩ $G = m_1\, m_2/(l_{1,2}\, F_{1,2})$
⟨8⟩ f
⟨9⟩ $G \approx 66.738\,4$ pN $\mathrm{m^2\ kg^{-2}}$; $u_c = 0.008\,0$ pN $\mathrm{m^2\ kg^{-2}}$

9.43U   **Units of dimension $\mathbf{L^3 M^{-1} T^{-2}}$**
⟨12⟩ *Coherent SI units* newton square metre(s) per kilogram squared or cubic metre(s) per kilogram second squared (N $\mathrm{m^2\ kg^{-2}} = \mathrm{m^3\ kg^{-1}\ s^{-2}}$)

9.44 Dimension $L^2 T^{-3}$

9.44.1 $\langle 2 \rangle$ **massic energy rate**
$\langle 3 \rangle$ **massic power**; specific power
$\langle 6 \rangle$ energy in a small time divided by mass of the system and by the time difference
$\langle 8 \rangle$ specific energy flux

9.44.1.1 $\langle 2 \rangle$ **massic energy rate of ionizing radiation** $\Delta\lambda$
$\langle 3 \rangle$ **absorbed dose rate**
$\langle 5 \rangle$ $\dot{D}$
$\langle 6 \rangle$ massic energy of an ionizing radiation, D, in a small time interval, t, divided by the time difference
$\langle 7 \rangle$ $\dot{D} = \mathrm{d}D/\mathrm{d}t$

9.44U **Units of dimension $L^2 T^{-3}$**
$\langle 12 \rangle$ *Coherent SI units of massic energy rate* watt(s) per kilogram or joule(s) per kilogram second or square metre(s) per second cubed $(\mathrm{W\ kg^{-1} = J\ kg^{-1}\ s^{-1} = m^2\ s^{-3}})$
$\langle 12 \rangle$ *Coherent SI units of massic energy rate of ionizing radiation* gray(s) per second or watt(s) per kilogram or joule(s) per kilogram second or square metre(s) per second cubed $(\mathrm{Gy\ s^{-1} = W\ kg^{-1} = m^2\ s^{-3}})$

9.45 Dimension MT^{-3}

9.45.1 $\langle 2 \rangle$ **areic energy rate**
$\langle 3 \rangle$ energy flux density; energy fluence rate
$\langle 6 \rangle$ energy in a small time divided by the area of the system and by the time difference
$\langle 8 \rangle$ energy flux; power density; flux density; fluence rate

9.45.1.1 $\langle 2 \rangle$ **areic heat rate** across surface or interface between subsystems 1 and 2
$\langle 3 \rangle$ areic heat flow rate
$\langle 5 \rangle$ $\varphi_{1,2}$
$\langle 6 \rangle$ heat rate, $\Phi_{1,2}$, divided by the area of surface or interface, $A_{1,2}$, across which heat is transferred
$\langle 7 \rangle$ $\varphi_{1,2} = \Phi_{1,2}/A_{1,2}$
$\langle 8 \rangle$ density of heat flow rate; $q_{1,2}$

9.45.1.2 $\langle 2 \rangle$ **areic energy rate of radiation** $\Delta\lambda$ **emitted** by a uniform body at uniform temperature or at a point on a surface
$\langle 3 \rangle$ **radiant exitance**; brightness; radiant emittance
$\langle 5 \rangle$ $M(\Delta\lambda)$, $M_{\mathrm{e}}(\Delta\lambda)$
$\langle 6 \rangle$ energy rate of the radiation emitted from a body, $P(\Delta\lambda)$, divided by its surface area, A
$\langle 7 \rangle$ $M_{\mathrm{e}}(\Delta\lambda) = P(\Delta\lambda)/A$
$\langle 8 \rangle$ radiant excitance
$\langle 9 \rangle$ (1) The term exitance has the root idea of *going out of* (or *exit*) and not of excitement of the source.
$\langle 9 \rangle$ (2) For the relationships of spectral quotients see §7.3.13.
$\langle 9 \rangle$ (3) For an unpolarized full radiator:

$$M_{\mathrm{e},\lambda} = c_0/(4\ w_{\mathrm{e},\lambda})$$
$$= 2\ \pi\ h\ c_0{}^2\ f(\lambda,T)$$

and

$$M_{\mathrm{e}} = \sigma\ T^4$$

where c_0 is the speed of radiation in vacuum, $w_{e,\lambda}$ is differential quotient of volumic energy rate to wavelength, h is the Planck constant, σ is the Stefan–Boltzmann constant, λ is wavelength and T is thermodynamic temperature.

9.45.1.3 $\langle 2 \rangle$ **areic energy rate of radiation $\Delta\lambda$ received** on surface 1
$\langle 3 \rangle$ **irradiance**
$\langle 5 \rangle$ $E_e(\Delta\lambda)$, $E(\Delta\lambda)$
$\langle 6 \rangle$ energy rate of a radiation received on a surface, $P(\Delta\lambda)$, divided by the area of that surface, A_1
$\langle 7 \rangle$ $E_e(\Delta\lambda) = P(\Delta\lambda)/A_1$
$\langle 9 \rangle$ The surface may be either a uniform surface or a point element of surface.

9.45.1.4 $\langle 2 \rangle$ **areic energy rate of a unidirectional sound wave** through a plane normal to the direction of propagation
$\langle 3 \rangle$ **sound intensity**
$\langle 5 \rangle$ I, J
$\langle 6 \rangle$ energy rate of a sound wave, P_a, divided by the area of a surface normal to the direction of propagation, A
$\langle 7 \rangle$ $I = P_a/A_\perp$

9.45U **Units of dimension MT^{-3}**
$\langle 12 \rangle$ *Coherent SI units* watt(s) per square metre or joule(s) per square metre second or kilogram(s) per second cubed ($W\,m^{-2} = J\,m^{-2}\,s^{-1} = kg\,s^{-3}$)

Term for unit	Symbol of unit Recommended	Other	Value in coherent SI unit
kilowatt(s) per square metre	$kW\,m^{-2}$		$= 1000\,W\,m^{-2}$
calorie(s) per square centimetre minute		$cal\,cm^{-2}\,min^{-1}$	$= 697.8\,W\,m^{-2}$
milliwatt(s) per square centimetre		$mW\,cm^{-2}$	$= 10\,W\,m^{-2}$
kilocalorie(s) per square metre hour		$kcal\,m^{-2}\,h^{-1}$	$= 1.163\,W\,m^{-2}$
kilogram(s) per second cubed	$kg\,m\,s^{-3}$		$= 1\,kg\,m\,s^{-3}$
watt(s) per square metre	$W\,m^{-2}$		$= 1\,W\,m^{-2}$
joule(s) per square metre second	$J\,m^{-2}\,s^{-1}$		$= 1\,W\,m^{-2}$
milliwatt(s) per square metre	$mW\,m^{-2}$		$= 10^{-3}\,W\,m^{-2}$
dyne(s) per square metre		$dyn\,m^{-2}$	$= 10^{-3}\,W\,m^{-2}$
erg(s) per square centimetre second		$erg\,cm^{-2}\,s^{-1}$	$= 10^{-3}\,W\,m^{-2}$
joule(s) per square metre day	$J\,m^{-2}\,d^{-1}$		$\approx 11.574\,07\,\mu W\,m^{-2}$
microwatt(s) per square metre	$\mu W\,m^{-2}$		$= 10^{-6}\,W\,m^{-2}$

9.46 Dimension $L^{-1}MT^{-3}$

9.46.1 $\langle 2 \rangle$ **volumic energy rate**
$\langle 5 \rangle$ w_e
$\langle 6 \rangle$ energy in a system, $Q_{e,1}$, in a small time, t, divided by volume of the system, V_1, and by the time difference
$\langle 7 \rangle$ $w_e = (dQ_{e,1}/dt)/V_1$
$\langle 8 \rangle$ power density; energy flux density

9.46.2   ⟨2⟩ **differential quotient of areic energy rate to wavelength** of radiation $\Delta\lambda$
⟨5⟩ $\varphi_{e,\lambda}$
⟨8⟩ spectral concentration of radiant flux density; spectral radiant flux density; spectral density of radiant flux density; spectral brightness
⟨9⟩ Definition is analogous to that of the differential quotient of energy rate to wavelength (§7.3.13 and §9.47.1).

9.46.3   ⟨2⟩ **differential quotient of steradic areic energy rate to wavelength** of radiation λ in direction ϑ, φ
⟨5⟩ $L_{e,\lambda}$, L_λ
⟨8⟩ spectral emissivity; spectral radiance
⟨9⟩ Definition is analogous to the differential quotient of energy rate to wavelength (§7.3.13 and §9.47.1).

9.46U   **Units of dimension $L^{-1}MT^{-3}$**
⟨12⟩ *Coherent SI units of volumic energy rate* watt(s) per cubic metre or kilogram(s) per metre second cubed (W m^{-3} = kg m^{-1} s^{-3})
⟨12⟩ *Coherent SI units of differential quotient of steradic areic energy rate to wavelength* watt(s) per cubic metre steradian or watt(s) per cubic metre or kilogram(s) per metre second cubed (W m^{-3} sr^{-1} = W m^{-3} = kg m^{-1} s^{-3})

9.47   Dimension LMT^{-3}

9.47.1   ⟨2⟩ **differential quotient of energy rate to wavelength** of radiation λ
⟨3⟩ spectral concentration of radiant power in terms of wavelength; spectral radiant power in terms of wavelength
⟨5⟩ $\Phi_{e,\lambda}$, P_λ
⟨6⟩ energy rate of a radiation, Φ_e, over a small wavelength interval, $\delta\lambda$, around a median wavelength, λ_0, divided by the wavelength difference
⟨7⟩ $\Phi_{e,\lambda}(\lambda_0) = \mathrm{d}\Phi_e/\mathrm{d}\lambda$
⟨8⟩ spectral concentration of radiant power; spectral radiant power; spectral density of radiant power
⟨9⟩ (1) The quantity may be estimated for the central wavelength of a small wavelength interval from the difference quotient $\Delta\Phi_e/\Delta\lambda$.
⟨9⟩ (2) The energy rate over a wavelength interval $[\lambda_1;\lambda_2]$ is the integral over wavelength of the differential quotient:

$$\Phi_e(\lambda_1 ; \lambda_2) = \int_1^2 \Phi_{e,\lambda}\mathrm{d}\lambda$$

⟨9⟩ (3) **Definition and practical usage are analogous to the differential quotient of energy rate to frequency (§9.42.3).**

9.47U   **Units of dimension LMT^{-3}**
⟨12⟩ *Coherent SI units* watt(s) per metre or kilogram metre(s) per second cubed (W m^{-1} = kg m s^{-3})

Term for unit	Symbol of unit		Value in coherent SI unit
	Recommended	Other	
watt(s) per ångström		W Å^{-1}	= 10×10^9 W m^{-1}
gigawatt(s) per metre	GW m^{-1}		= 10^9 W m^{-1}
watt(s) per nanometre		W nm^{-1}	= 10^9 W m^{-1}
megawatt(s) per metre	MW m^{-1}		= 10^6 W m^{-1}
watt(s) per micrometre		W μm^{-1}	= 10^6 W m^{-1}
kilowatt(s) per metre	kW m^{-1}		= 1000 W m^{-1}
watt(s) per millimetre		W mm^{-1}	= 1000 W m^{-1}
watt(s) per metre	W m^{-1}		= 1 W m^{-1}

⟨19⟩ Units such as watt per nanometre are accepted in spectrometry.

9.48 Dimension L^2MT^{-3}

9.48.1   ⟨2⟩ **energy rate**
⟨3⟩ **power**
⟨5⟩ P, Φ_e
⟨6⟩ energy transferred or transformed, Q_e, in a small time interval, t, divided by the time difference
⟨7⟩ $P = dQ_e/dt$
⟨8⟩ energy flux
⟨9⟩ The energy rate of electricity is the product of electrical current, I, and electric potential difference, U:

$$P = I \times U$$

9.48.1.1   ⟨2⟩ **heat rate** across surface or interface between subsystems 1 and 2
⟨3⟩ **heat flow rate**
⟨5⟩ $\Phi_{1,2}$, P
⟨6⟩ amount-of-heat transferred, H, in a small time interval, t, divided by the time difference
⟨7⟩ $\Phi_{1,2} = dH/dt$

9.48.1.2   ⟨2⟩ **energy rate of radiation** $\Delta\lambda$ into or from system 1
⟨3⟩ **radiant flux; radiant power; radiant energy flux**
⟨5⟩ $\Phi_e(\Delta\lambda)$, $P(\Delta\lambda)$
⟨6⟩ energy of the radiation emitted, transferred or received to or from the system, $Q_e(\Delta\lambda)$, in a small time interval, t, divided by the time difference
⟨7⟩ $\Phi_e(\lambda) = dQ_e(\lambda)/dt$
⟨9⟩ See §7.3.5 and §7.3.9 for its application in optical spectroscopy.

9.48.2   ⟨2⟩ **steradic energy rate of radiation** λ from source 1 in direction ϑ, φ
⟨3⟩ **radiant intensity**
⟨5⟩ I_e, I
⟨6⟩ energy rate of a radiation, $P(\lambda)$, in given direction, ϑ, ϕ from a source in a small solid angle, Ω, divided by that solid angle
⟨7⟩ $I_{e,\lambda} = [dP_e(\lambda)/d\Omega]_{\vartheta,\varphi}$
⟨8⟩ intensity

9.48U   **Units of dimension L^2MT^{-3}**
⟨12⟩ *Coherent SI units of energy rate* watt(s) or joule(s) per second or volt(s) per ampere or kilogram metre(s) squared per second cubed (W = J s^{-1} = V A^{-1} = kg m^2 s^{-3})
⟨12⟩ *Coherent SI units of steradic energy rate of radiation* watt(s) per steradian or watt(s) or kilogram metre(s) squared per second cubed (W sr = W = kg m^2 s^{-3})

Term for unit	Recommended	Other	Value in coherent SI unit
megawatt(s)	MW		$= 10^6$ W
kilowatt(s)	kW		$= 1000$ W
megajoule(s) per day	MJ d^{-1}		$\approx 11.574\,07$ W
kilogram-force metre(s) per second		kgf m s^{-1}	$= 9.806\,65$ W
calorie(s) per second		cal s^{-1}	$= 4.186\,8$ W
watt(s)	W		$= 1$ W
joule(s) per second	J s^{-1}		$= 1$ W
calorie(s) per minute		cal min^{-1}	$\approx 69.78 \times 10^{-3}$ W

(*Continued*)

Term for unit	Symbol of unit		Value in coherent
	Recommended	Other	SI unit
calorie(s) per hour		cal h^{-1}	$\approx 1.163 \times 10^{-6}$ W
milliwatt(s)	mW		$= 10^{-3}$ W
calorie(s) per day		cal$_{th}$ d^{-1}	$\approx 48.425\,93 \times 10^{-6}$ W
microwatt(s)	μW		$= 10^{-6}$ W
erg(s) per second		erg s^{-1}	$= 100 \times 10^{-9}$ W
nanowatt(s)	nW		$= 10^{-9}$ W

9.49 Dimension L^{3}MT^{-3}

9.49.1   ⟨2⟩ **differential quotient of energy rate to wavenumber** of radiation ν
⟨3⟩ spectral concentration of radiant power in terms of wavenumber; spectral radiant power in terms of wavenumber
⟨5⟩ $\Phi_{e,\tilde{\nu}}$, $P_{\tilde{\nu}}$
⟨8⟩ spectral concentration of radiant power; spectral radiant power; spectral density of radiant power
⟨9⟩ Definition and practical use are analogous to the differential quotient of energy rate to wavelength (§7.3.13 and §9.47.1).

9.49U   **Units of dimension L^{3}MT^{-3}**
⟨12⟩ *Coherent SI units* watt(s) per metre or kilogram cubic metre(s) per second cubed (W m^{-1} = kg m^3 s^{-3})

9.50 Dimension TI

9.50.1   ⟨2⟩ **electrical charge**
⟨3⟩ quantity of electricity; electrical amount; amount-of-electricity; electric charge
⟨5⟩ Q
⟨6⟩ integral of electric current, I, over time, t
⟨7⟩ $Q = \int I\,dt$
⟨8⟩ charge
⟨9⟩ The term electric (without "al") charge which is currently used does not agree with the recommendation of this Compendium (§5.4.6.14).

9.50.2   ⟨2⟩ **electrical charge constant**
⟨3⟩ **elementary charge**
⟨5⟩ e
⟨6⟩ electrical charge of a proton
⟨8⟩ charge
⟨9⟩ $e = 0.160\,217\,656\,5$ aC; $u_c = 0.000\,000\,003\,5$ aC
The assignment of an exact value to the elementary charge is under discussion (CIPM, CCU), see §5.1.9.

9.50U   **Units of dimension TI**
⟨12⟩ *Coherent SI units of electrical charge* coulomb(s) or ampere second(s) (C = A s)

9.51 Dimension M^{-1}TI

9.51.1   ⟨2⟩ **massic electrical charge**
⟨5⟩ q
⟨6⟩ electrical charge in a system, Q_1, divided by its mass, m_1
⟨7⟩ $q_1 = Q_1/m_1$

9.51.1.1 ⟨2⟩ **massic electrical charge induced by ionizing radiation** λ in air
⟨5⟩ *X*
⟨6⟩ electrical charge in an element of air, Q_{air}, due to ions of the same sign induced when all the electrons liberated by photons of a specified radiation are stopped in that air, divided by the mass of that air, m_{air}
⟨7⟩ $X(\lambda) = Q_{air}/m_{air}$
⟨8⟩ exposure

9.51U **Units of dimension M^{-1}TI**
⟨12⟩ *Coherent SI units* coulomb(s) per kilogram or ampere(s) second per kilogram (C kg^{-1} = A s kg^{-1})

Term for unit	Symbol of unit		Value in coherent SI unit
	Recommended	Other	
coulomb(s) per kilogram	C kg^{-1}		= 1 C kg^{-1}
ampere(s) second per kilogram	A s kg^{-1}		= 1 C kg^{-1}
millicoulomb(s) per kilogram	mC kg^{-1}		= 0.001 C kg^{-1}
roentgen(s)		R	= 0.258 × 10^{-3} C kg^{-1}

9.52 Dimension I

9.52.1 ⟨2⟩ **electrical current**
⟨3⟩ electricity rate, electric current
⟨5⟩ *I, i*
⟨8⟩ current
⟨9⟩ (1) Electrical current equals the electrical charge transferred, *Q*, in a small time interval, *t*, divided by the time difference:

$$I = dQ/dt$$

⟨9⟩ (2) The symbol *I* may be subscripted a for anodic, c for cathodic, e or 0 for exchange, and l for limiting.
⟨10⟩ Interface(anode, electrolyte)—Electrons; electrical current = ? A
⟨9⟩ (3) Anodic partial electrical current is also used in place of electrical current; I_a

9.52U **Units of dimension I**
⟨12⟩ *SI base unit* ampere(s) or (coulomb(s) per second) (A = C s^{-1})
⟨16⟩ The ampere is that constant current which, if maintained in two straight parallel conductors of infinite length, of negligible circular cross-section, and placed 1 metre apart in vacuum, would produce between these conductors a force equal to 2 × 10^{-7} newton per metre of length (CIPM 1946, Resolution 2; 9th CGPM 1948). An alternative definition of the ampere is under discussion (CIPM, CCU), see §5.1.9.

9.53 Dimension L^{-1}I

9.53.1 ⟨2⟩ **magnetic field strength**
⟨5⟩ *H* (as scalar), *H* (as vector), $\vec{H}$ (as vector)
⟨6⟩ the driving magnetic influence from external currents in a material
⟨7⟩ $H = Xj$ where *R* is reactance *i.e.* imaginary part of the impedance.

9.53U **Units of dimension L^{-1}I**
⟨12⟩ *Coherent SI units* ampere(s) per metre or newton(s) per weber or coulomb(s) per metre second (A m^{-1} = N Wb^{-1} = C m^{-1} s^{-1})

9.54 Dimension $L^{-2}I$

9.54.1 $\langle 2 \rangle$ **areic electrical current** of electrode 1
$\langle 3 \rangle$ areic electricity rate; electric current density
$\langle 5 \rangle$ j, J
$\langle 6 \rangle$ electrical current across the surface of an electrode, I_1, divided by the area of that electrode, $A_{\perp,1}$
$\langle 7 \rangle$ $J_1 = I_1/A_{\perp,1}$
$\langle 8 \rangle$ current density

9.54U **Units of dimension $L^{-2}I$**
$\langle 12 \rangle$ *Coherent SI units* ampere(s) per square metre or coulomb(s) per square metre second (A m^{-2} = C m^{-2} s)

9.55 Dimension $M^{-1}T^2I$

9.55.1 $\langle 2 \rangle$ **electrophoretic mobility** of ionic component B
$\langle 3 \rangle$ **electrical mobility; electrolytic mobility**
$\langle 5 \rangle$ μ_B, m
$\langle 6 \rangle$ distance travelled or migrated (also termed drift) of a component, l_B, in a time interval, t, in an electric field divided by the time difference and by the electrical field strength, E, in a given medium
$\langle 7 \rangle$ $\mu_B = (\Delta l_B/\Delta t)/E$
$\qquad = v_B/E$
$\langle 8 \rangle$ mobility
$\langle 9 \rangle$ (1) Values are negative if migration is against the electric field.
$\langle 9 \rangle$ (2) For an example of calculation of electrophoretic mobility see §7.5.2.

9.55U **Units of dimension $M^{-1}T^2I$**
$\langle 12 \rangle$ *Coherent SI units* square metre(s) per second volt or coulomb second(s) per kilogram or ampere(s) second squared per kilogram
$\left(m^2 \ s^{-1} \ V^{-1} = C \ s \ kg^{-1} = A \ s^2 \ kg^{-1} \right)$

9.56 Dimension $MT^{-2}I^{-1}$

9.56.1 $\langle 2 \rangle$ **magnetic induction**
$\langle 3 \rangle$ magnetic flux density
$\langle 5 \rangle$ B
$\langle 6 \rangle$ vector kind-of-quantity such that the force exerted on an element of electricity rate, F, equals the vector product of that element, I, and the magnetic induction
$\langle 7 \rangle$ $F = I \, \Delta s \times B$

9.56U **Units of dimension $MT^{-2}I^{-1}$**
$\langle 12 \rangle$ *Coherent SI units* of magnetic induction tesla(s) or kilogram(s) per second squared ampere, or kilogram(s) per second coulomb (T = kg s^{-2} A^{-1} = kg s^{-1} C^{-1})

9.57 Dimension $L^2MT^{-2}I^{-1}$

9.57.1 $\langle 2 \rangle$ **magnetic flux**
$\langle 5 \rangle$ Φ_m
$\langle 6 \rangle$ scalar product of the magnetic induction across a surface element, B, and the area of the surface element, A
$\langle 7 \rangle$ $\Phi_m = \int B \times e \times dA$ where e is the unit vector.

9.57U   Units of dimension $L^2MT^{-2}I^{-1}$
⟨12⟩ *Coherent SI units* weber(s) or tesla square metre(s) or volt second(s) or newton tesla(s) per ampere square metre or kilogram square metre(s) per second squared ampere or kilogram square metre(s) per second coulomb
$(Wb = T\ m^2 = V\ s = N\ T\ A^{-1}\ m^{-2} = kg\ m^2\ s^{-2}\ A^{-1} = kg\ m^2\ s^{-1}\ C^{-1})$

9.58   Dimension $LMT^{-3}I^{-1}$

9.58.1   ⟨2⟩ **electric field strength**
⟨5⟩ E (as scalar), E (as vector)
⟨6⟩ force exerted by an electric field on an electrical point charge, F, divided by the electrical charge, Q
⟨7⟩ $E = F/Q$
⟨9⟩ In electrophoresis, the counterforce exerted by the point charge is symbolized F':

$$F' = 6\ \pi\ r_B\ \eta_1\ v_B$$

where r_B is the radius of the ionic particle, η_1 is the dynamic viscosity of the medium and v_B is the velocity of the particle.

9.58U   Units of dimension $LMT^{-3}I^{-1}$
⟨12⟩ *Coherent SI units* volt(s) per metre or newton(s) per coulomb or kilogram metre(s) per ampere second cubed or kilogram metre(s) per coulomb second squared $(V\ m^{-1} = N\ C^{-1} = kg\ m\ A^{-1}\ s^{-3} = kg\ m\ C^{-1}\ s^{-2})$

9.59   Dimension $L^2MT^{-3}I^{-1}$

9.59.1   ⟨2⟩ **electric potential** for an electrostatic field in system 1
⟨5⟩ V, φ
⟨6⟩ (1) scalar value of the gradient of electric field strength, E
⟨7⟩ $V = |\nabla E|^{-1}$
⟨6⟩ (2) change in internal energy of a system U_1, that occurs when there is a change in the electrical charge of the system, Q_1
⟨7⟩ (2) $V_1 = (dU_1/dQ_1)_{V,S,n(A),n(B),\ldots,n(N)}$
⟨8⟩ potential
⟨9⟩ The symbol φ is preferred for inner electric potential.

9.59.2   ⟨2⟩ **electric potential difference** of electrodes 1 and 2 of an electrolytic cell
⟨3⟩ **electric tension**
⟨4⟩ p.d., PD
⟨5⟩ U, ΔV
⟨6⟩ difference in electric potential, V, between the two electrodes of a cell
⟨7⟩ $U = V_2 - V_1$
⟨8⟩ voltage; potential difference; tension
⟨9⟩ Electrode potential with respect to a cation-reversible reference electrode is represented as E_+ and that with respect to an anion-reversible reference electrode as E_-.

9.59.3   ⟨2⟩ **electromotive force** of source 1
⟨5⟩ E, E_{mf}
⟨6⟩ energy supplied by a source divided by the electrical charge transported through the source
⟨7⟩ $E = (F/Q)\ dr$ where F is force, Q is electric charge, and r is internal resistance.
⟨8⟩ EMF, e.m.f.
⟨9⟩ The term "electromotive force", and the symbol "e.m.f." are not recommended because an electric potential difference is not a force.[13]

9.59.4 ⟨2⟩ **electrokinetic potential** of solution 1
⟨3⟩ zeta potential
⟨5⟩ ζ
⟨6⟩ electric potential difference between the fixed charges of an immobile support and the diffuse charge in a solution

9.59U **Units of dimension $L^2MT^{-3}I^{-1}$**
⟨12⟩ *Coherent SI units* volt(s) or joule(s) per coulomb or joule(s) per ampere second or kilogram square metre(s) per ampere second cubed or kilogram square metre(s) per coulomb second squared ($V = J\ C^{-1} = J\ A^{-1}\ s^{-1} = kg\ m^2\ A^{-1}\ s^{-3} = kg\ m^2\ C^{-1}\ s^{-2}$)

9.60 Dimension $L^2MT^{-2}I^{-2}$

9.60.1 ⟨2⟩ **self-inductance** of a thin conducting loop
⟨5⟩ L
⟨6⟩ magnetic flux through a loop, Φ_m, due to an electrical current in the loop, I, divided by that electrical current
⟨7⟩ $L = \Phi_m/I$

9.60U **Units of dimension $L^2MT^{-2}I^{-2}$**
⟨12⟩ *Coherent SI units* henry (pl. henries) or weber(s) per ampere or kilogram square metre(s) per ampere squared second squared or kilogram square metre(s) per coulomb squared ($H = Wb\ A^{-1} = kg\ m^2\ A^{-2}\ s^{-2} = kg\ m^2\ C^{-2}$)

9.61 Dimension $L^{-2}M^{-1}T^3I^2$

9.61.1 ⟨2⟩ **electrical conductance** of conductor 1
⟨3⟩ electric conductance
⟨5⟩ G
⟨6⟩ reciprocal of the electric resistance of a conductor, R_1
⟨7⟩ $G_1 = 1/R_1$
⟨8⟩ conductance

9.61U **Units of dimension $L^{-2}M^{-1}T^3I^2$**
⟨12⟩ *Coherent SI units* siemens(es) or (one) per ohm or reciprocal ohm(s) or ampere(s) per volt or ampere(s) squared per watt or ampere(s) squared second cubed per kilogram squared metre or second coulomb(s) squared per kilogram square metre ($S = \Omega^{-1} = A\ V^{-1} = A^2\ W^{-1} = s^3\ A^2\ kg^{-1}\ m^2 = s\ C^2\ kg^{-1}\ m^2$)

9.62 Dimension $L^2MT^{-3}I^{-2}$

9.62.1 ⟨2⟩ **electric resistance** of conductor 1
⟨5⟩ R
⟨6⟩ electric potential difference, U_1, divided by electrical current, I_1, when there is no electromotive force
⟨7⟩ $R_1 = (U_1/I_1)_{E=0}$
⟨8⟩ resistance
⟨9⟩ For alternating current, electric resistance is defined as the real part of impedance (§9.62.2).

9.62.2 ⟨2⟩ **impedance**
⟨3⟩ **complex impedance**
⟨5⟩ Z
⟨6⟩ electric potential difference (§9.59.2) divided by electrical current (§9.52.1) and X is reactance
⟨7⟩ $Z = R + jX$ where R is electric resistance (§9.62.1) and X is reactance.

9.62U   Units of dimension $L^2MT^{-3}I^{-2}$

$\langle 12 \rangle$ *Coherent SI units* ohms or (one) per siemens or reciprocal siemens(es) or volt(s) per ampere or watt(s) per ampere squared or kilogram square metre(s) per second cubed per ampere squared or kilogram square metre(s) per second coulomb squared $(\Omega = S^{-1} = V\,A^{-1} = W\,A^{-2} = kg\,m^2\,s^{-3}\,A^{-2} = kg\,m^2\,s^{-1}\,C^{-2})$

9.63   Dimension $L^{-3}M^{-1}T^3I^2$

9.63.1 $\langle 2 \rangle$ **lineic electrical conductance** of conductor 1

$\langle 3 \rangle$ **electrical conductivity**; electrolytic conductivity

$\langle 5 \rangle$ γ, σ, κ

$\langle 6 \rangle$ (1) reciprocal of the electric resistivity of a conductor, ρ_1 (§9.64.1)

$\langle 7 \rangle$ $\gamma = 1/\rho$

$\langle 6 \rangle$ (2) areic electrical current through an electrolyte (or other conductor), j, divided by the electric field strength, F

$\langle 7 \rangle$ $\gamma = j/F$

$\langle 8 \rangle$ conductivity; specific conductance; specific conductivity

9.63U   Units of dimension $L^{-3}M^{-1}T^3I^2$

$\langle 12 \rangle$ *Coherent SI units* siemens(es) per metre or ampere squared second(s) cubed per kilogram cubic metre, or second coulomb(s) squared per kilogram cubic metre $(S\,m^{-1} = A^2\,s^3\,kg^{-1}\,m^{-3} = s\,C^2\,kg^{-1}\,m^{-3})$

9.64   Dimension $L^3MT^{-3}I^{-2}$

9.64.1 $\langle 2 \rangle$ **electric resistivity** of conductor 1

$\langle 5 \rangle$ ρ_1

$\langle 6 \rangle$ electric field strength, E_1, divided by electrical current density, J_1, when there is no electromotive force

$\langle 7 \rangle$ $\rho_1 = (E_1/J_1)_{E=0}$

$\langle 8 \rangle$ resistivity

9.64U   Units of dimension $L^3MT^{-3}I^{-2}$

$\langle 12 \rangle$ *Coherent SI units* ohm metre(s) or kilogram cubic metre(s) per second coulomb squared or kilogram cubic metre(s) per ampere squared second cubed $(\Omega\,m = kg\,m^3\,s^{-1}\,C^{-2} = kg\,m^3\,A^{-2}\,s^{-3})$

9.65   Dimension $M^{-1}T^4I^2$

9.65.1 $\langle 2 \rangle$ **electrical polarizability** of molecule B in an electrical field

$\langle 5 \rangle$ α, γ

$\langle 6 \rangle$ electrical dipole moment of the molecule induced by an electrical field divided by electric field strength

9.65U   Units of dimension $M^{-1}T^4I^2$

$\langle 12 \rangle$ *Coherent SI units* coulomb square metre(s) per volt or ampere squared second(s) to the power four per kilogram, or coulomb second(s) squared per kilogram $(C\,m^2\,V^{-1} = A^2\,s^4\,kg^{-1} = C^2\,s^2\,kg^{-1})$

9.66   Dimension $L^{-2}M^{-1}T^4I^2$

9.66.1 $\langle 2 \rangle$ **electrical capacitance** of capacitor 1

$\langle 3 \rangle$ **capacitance**; electric capacitance

$\langle 5 \rangle$ C

⟨6⟩ electrical charge in a capacitor, Q_1, divided by its electric potential difference, U_1

⟨7⟩ $C_1 = Q_1/U_1$

9.66U   Units of dimension $L^{-2}M^{-1}T^4I^2$

⟨12⟩ *Coherent SI units* farad(s) or coulomb(s) squared per joule or ampere squared second(s) to the power four per kilogram square metre or coulomb squared second(s) squared per kilogram square metre

$(F = C^2 J^{-1} = s^4 A^2 kg^{-1} m^{-2} = s^2 C^2 kg^{-1} m^{-2})$

9.67   Dimension Θ

9.67.1   ⟨2⟩ **thermodynamic temperature** of system 1

⟨3⟩ **Kelvin temperature**; absolute temperature

⟨5⟩ T_1, θ_1

⟨8⟩ temperature

⟨9⟩ (1) In routine, the kind-of-quantity Celsius temperature is used.

⟨9⟩ (2) The term "centigrade temperature" is discouraged.

9.67.1.1 ⟨2⟩ **temperature difference**

⟨4⟩ temp.diff.

⟨5⟩ $\Delta\vartheta$, $\Delta\theta$, ΔT

⟨8⟩ temperature interval

⟨9⟩ Because the units of thermodynamic and Celsius temperature are equal, a difference of temperature can be expressed either in kelvins or in degrees Celsius.

⟨10⟩ P—Freezing point depression; temp.diff.(reference material water) = 545 mK

9.67.2   ⟨2⟩ **Celsius temperature** of system 1

⟨5⟩ θ_1, t_1

⟨7⟩ $\theta_1 = T_1 - T_0$

⟨8⟩ temperature

⟨9⟩ (1) The symbol θ is preferred over t to avoid confusion with t for time. The symbol T should be reserved for thermodynamic temperature (§9.67.1).

⟨9⟩ (2) Celsius temperature t is related to thermodynamic temperature T by

$$t/°C = T/K - 273.15$$

⟨9⟩ (3) For Fahrenheit temperature:

$$T_0 = 459.67 \ °R \quad \text{where °R indicates degree Rankine.}$$

$$= [(5 \times 459.67)/9] \ K$$

⟨9⟩ (4) For Celsius temperature:

$$T_0 = 273.15 \ K$$

$$= T(\text{triple point}) - 0.01 \ K$$

⟨9⟩ (5) The International Temperature Scale of 1990, ITS-90, replaces the International Practical Temperature Scale of 1968, IPTS-68, and is based on various fixed points and interpolation procedures. The kinds-of-quantity symbols are represented by the subscript 90:

$$t_{90} = T_{90} - T_0 \ [\text{ref. 2}]$$

Celsius temperature, t_{90}, and thermodynamic temperature, T_{90}, are thus defined by the same series of reference values; International Celsius temperature is no longer defined by the freezing and boiling points of water. Uncertainties have been reduced to 1 or 2 millikelvins up to thermodynamic temperatures of about 373 K but increase progressively above that value.

⟨10⟩ Air(Room)—Wet bulb; Celsius temp. = 22.5 °C
⟨10⟩ Pt(spec.)—Blood; Celsius temp. = 37.9 °C
⟨10⟩ Pt—Body; Celsius temp. = 36.8 °C
⟨10⟩ Room—Air(saturated with water vapour); Celsius temp. = 18.5 °C

9.67U   Units of dimension Θ

⟨12⟩ *SI base unit of thermodynamic temperature* kelvin(s) (K)
⟨12⟩ *SI unit of Celsius temperature* degree(s) Celsius (°C = K)
⟨16⟩ The kelvin, unit of thermodynamic temperature, is the fraction 1/273.16 of the thermodynamic temperature of the triple point of water (13th CGPM 1967, Resolution 3). So, Celsius temperature at 0 °C is 273 K (see ISO/IEC 80000-5 [14]). This definition refers to water having the isotopic composition defined exactly by the following amount-of-substance ratios: 0.000 155 76 mole of ^{2}H per mole of ^{1}H, 0.000 379 9 mole of ^{17}O per mole of ^{16}O, and 0.002 000 52 mole of ^{18}O per mole of ^{16}O (CIPM 2005). An alternative definition of kelvin is under discussion (CIPM, CCU), see §5.1.9.

Term for unit	Symbol of unit		Value in coherent SI unit
	Recommended	Other	
kelvin(s)	K		= 1 K
degree(s) Celsius	°C		= 1 °C = 1 K
degree(s) Fahrenheit		°F	= (5/9) °C
millikelvin(s)	mK		= 0.001 K
millidegree(s)		mdeg	= 0.001 K

⟨19⟩ (1) In contrast to usage for all other kinds-of-quantity, choice of the unit of temperature implies a defined zero point.
⟨19⟩ (2) The symbol °C is a single symbol; the degree sign should not be spaced from the letter C; a space is mandatory between the numerical value and the unit symbol, 37 °C.

9.68   Dimension Θ^{-1}

9.68.1   ⟨2⟩ **expansion coefficient**
⟨5⟩ α

9.68.1.1   ⟨2⟩ **linear expansion coefficient**
⟨5⟩ α_l
⟨7⟩ $\alpha_l = \mathrm{d}\ln l/\mathrm{d}T$
$= (\mathrm{d}l/\mathrm{d}T)/l$
⟨8⟩ expansion coefficient; coefficient of linear thermal expansion

9.68.1.2   ⟨2⟩ **cubic expansion coefficient**
⟨5⟩ α_V
⟨7⟩ $\alpha_V = \mathrm{d}\ln V/\mathrm{d}T$
$= (\mathrm{d}V/\mathrm{d}T)/V$
⟨8⟩ expansion coefficient; volumetric coefficient

9.68.1.3   ⟨2⟩ **relative pressure coefficient**
⟨5⟩ α_p
⟨7⟩ $\alpha_p = \mathrm{d}\ln p/\mathrm{d}T$
$= (\mathrm{d}p/\mathrm{d}T)/p$
⟨8⟩ pressure coefficient

9.68U **Units of dimension Θ^{-1}**
⟨12⟩ *Coherent SI units* one per kelvin or per kelvin or reciprocal kelvin(s) or (one) per degree Celsius or reciprocal degree(s) Celsius
$(K^{-1} = C^{-1})$

9.69 Dimension $L^{-1}\Theta$

9.69.1 ⟨2⟩ **temperature gradient**
⟨5⟩ **grad** T, ∇T
⟨7⟩ $\nabla T = dT/(\partial x, \partial y, \partial z)$

9.69U **Unit of dimension $L^{-1}\Theta$**
⟨12⟩ *Coherent SI units* kelvin(s) per metre $(K\ m^{-1})$ or degree(s) Celsius per metre $(^\circ C\ m^{-1})$

9.70 Dimension $T^{-1}\Theta$

9.70.1 ⟨2⟩ **temperature rate** of system 1
⟨5⟩ $\dot{T}_1$
⟨7⟩ $\dot{T}_1 = dT_1/dt$
⟨10⟩ Pt—Temperature rate(decrease, proc.) = 1.8 mK s^{-1}

9.70U **Unit of dimension $T^{-1}\Theta$**
⟨12⟩ *Coherent SI units* kelvin(s) per second $(K\ s^{-1})$ or degree(s) Celsius per second $(^\circ C\ s^{-1})$

9.71 Dimension $L^2T^{-2}\Theta^{-1}$

9.71.1 ⟨2⟩ **massic kelvic enthalpy** of system 1
⟨3⟩ **massic heat capacity**
⟨5⟩ c
⟨6⟩ kelvic enthalpy (or heat capacity) of a system, C_1, divided by its mass, m_1
⟨7⟩ $c_1 = C_1/m_1$
⟨8⟩ specific heat capacity; specific heat
⟨9⟩ Massic kelvic enthalpy at a specified pressure and a specified volume are indicated by subscripts p and V, respectively.

9.71.2 ⟨2⟩ **massic entropy** of system 1
⟨5⟩ s
⟨6⟩ entropy of a system, S_1, divided by its mass, m_1
⟨7⟩ $s_1 = S_1/m_1$
⟨8⟩ specific entropy

9.71U **Units of dimension $L^2T^{-2}\Theta^{-1}$**
⟨12⟩ *Coherent SI units* joule(s) per kilogram kelvin or joule(s) per kilogram degree Celsius, or square metre(s) per second squared kelvin
$(J\ kg^{-1}\ K^{-1} = J\ kg^{-1}\ {}^\circ C^{-1} = m^2\ s^{-2}\ K^{-1})$

Term for unit	Symbol of unit Recommended	Other	Value in coherent SI unit
calorie(s) per gram degree Celsius		cal $g^{-1}\ {}^\circ C^{-1}$	$= 4.186\ 8 \times 10^3\ J\ kg^{-1}\ K^{-1}$
kilojoule(s) per kilogram kelvin	kJ $kg^{-1}\ K^{-1}$		$= 1000\ J\ kg^{-1}\ K^{-1}$

9.72 Dimension $L^{-1}MT^{-2}\Theta^{-1}$

9.72.1 ⟨2⟩ **kelvic pressure** of gaseous system 1
⟨5⟩ β
⟨7⟩ $\beta_1 = dp_1/dT_1$
⟨8⟩ pressure coefficient

9.72U **Units of dimension $L^{-1}MT^{-2}\Theta^{-1}$**
⟨12⟩ *Coherent SI units* pascal(s) per kelvin or kilogram(s) per metre second squared kelvin ($Pa\ K^{-1} = kg\ m^{-1}\ s^{-2}\ K^{-1}$)

9.73 Dimension $L^2MT^{-2}\Theta^{-1}$

9.73.1 ⟨2⟩ **kelvic enthalpy** of system 1
⟨3⟩ **heat capacity**
⟨5⟩ C
⟨6⟩ increase in amount-of-heat in a system, Q_1, divided by the increase in its thermodynamic temperature, T_1
⟨7⟩ $C_1 = dQ_1/dT_1$
⟨9⟩ The type of change (*e.g.* constant pressure or constant volume) must be specified.

9.73.2 ⟨2⟩ **entitic kelvic energy constant**
⟨3⟩ **Boltzmann constant; molecular gas constant; entitic gas constant**
⟨5⟩ k, k_B
⟨6⟩ molar kelvic energy constant (or molar gas constant), R, divided by the Avogadro constant), N_A
⟨7⟩ $k = R/N_A$
$\approx 13.806\ 504$ yJ K^{-1}; $u_c = 0.000\ 024$ yJ K^{-1}
⟨9⟩ The assignment of a numerical value to the Boltzmann constant is under discussion (CIPM, CCU), see §5.1.9.

9.73.3 ⟨2⟩ **entropy** of system 1
⟨5⟩ S
⟨6⟩ (1) increase in the amount-of-heat of a system, Q_1, divided by its thermodynamic temperature, T_1, provided that there is no irreversible change in the system
⟨7⟩ (1) $dS_1 = dQ_1/T_1$
⟨7⟩ (2) $S_1 = \int T_1^{-1}\ dU_{1, dV = 0,\ dQ = 0,\ dn(B) = 0,\ \dots\ dn(N) = 0}$
⟨9⟩ Only a change in entropy can be measured.

9.73U **Units of dimension $L^2MT^{-2}\Theta^{-1}$**
⟨12⟩ *Coherent SI units* joule(s) per kelvin or kilogram square metre(s) per second squared kelvin ($J\ K^{-1} = kg\ m^2\ s^{-2}\ K^{-1}$)

Term for unit or constant	Symbol of unit or constant	Value in coherent SI unit
	Recommended	
kilojoule(s) per kelvin	kJ K^{-1}	$= 1000$ J K^{-1}
kilogram square metre(s) per second squared kelvin		$= kg\ m^2\ s^2\ K^{-1}$
Boltzmann constant	k	$\approx 13.806\ 488 \times 10^{-24}$ J K^{-1}
		$u_c = 0.000\ 013 \times 10^{-24}$ J K^{-1}
yoctojoule(s) per kelvin	yJ K^{-1}	$= 10^{-24}$ J K^{-1}

9.74 Dimension $M^{-1}T^3\Theta$

9.74.1 $\langle 2 \rangle$ **insulation coefficient** of interface between subsystems 1 and 2
$\langle 3 \rangle$ **insulance; thermal insulance; coefficient of thermal insulation**
$\langle 5 \rangle$ M, I
$\langle 6 \rangle$ (1) temperature difference across an interface, $\Delta_1^2 T$, divided by the areic heat rate across it, $\varphi_{1,2}$
$\langle 7 \rangle$ (1) $M_{1,2} = \Delta_1^2 T/\varphi_{1,2}$
$\langle 7 \rangle$ (2) $M_{1,2} = 1/K_{1,2}$ where $K_{1,2}$ is kelvic areic heat rate (§9.75.1).
$\langle 8 \rangle$ thermal resistance; R

9.74U **Units of dimension $M^{-1}T^3\Theta$**
$\langle 12 \rangle$ *Coherent SI units* kelvin square metre(s) per watt or kelvin second(s) cubed per kilogram (K m^2 W^{-1} = K s^3 kg^{-1})

9.75 Dimension $MT^{-3}\Theta^{-1}$

9.75.1 $\langle 2 \rangle$ **kelvic areic heat rate**
$\langle 5 \rangle$ K, α

9.75.1.1 $\langle 2 \rangle$ **kelvic areic heat rate across interface** of subsystems 1 and 2
$\langle 3 \rangle$ **heat transfer coefficient; coefficient of heat transfer**
$\langle 5 \rangle$ K, α
$\langle 6 \rangle$ areic heat rate across an interface, $\varphi_{1,2}$, divided by the temperature difference, $\Delta_1^2 T$
$\langle 7 \rangle$ $K_{1,2} = \varphi_{1,2}/\Delta_1^2 T$
$\langle 8 \rangle$ k

9.75U **Units of dimension $MT^{-3}\Theta^{-1}$**
$\langle 12 \rangle$ *Coherent SI units* watt(s) per square metre kelvin or kilogram(s) per second cubed kelvin (W m^{-2} K^{-1} = kg s^{-3} K^{-1})

Term for unit	Symbol of unit		Value in coherent SI unit
	Recommended	Other	
kilowatt(s) per square metre kelvin	kW m^{-2} K^{-1}		= 1000 W m^{-2} K^{-1}
kilocalorie(s) per square metre hour degree Celsius		kcal m^{-2} h^{-1} °C^{-1}	= 1.163 W m^{-2} K^{-1}
watt(s) per square metre kelvin	W m^{-2} K^{-1}		= 1 W m^{-2} K^{-1}
kilogram(s) per second cubed kelvin			= kg s^{-3} K^{-1}

9.76 Dimension $LMT^{-3}\Theta^{-1}$

9.76.1 $\langle 2 \rangle$ **lineic kelvic heat rate** of conducting system 1 in direction x
$\langle 3 \rangle$ **thermal conductivity; thermal conduction coefficient**
$\langle 5 \rangle$ $\lambda_{1,x}, \kappa$
$\langle 6 \rangle$ areic heat rate in a direction, φ_h, divided by the temperature gradient in that direction, $\nabla_x T_1$
$\langle 7 \rangle$ $\lambda_{1,x} = \varphi_h/\nabla_x T_1$
$\langle 9 \rangle$ Lineic kelvic heat rate is a vector kind-of-quantity.

9.76U **Units of dimension LMT$^{-3}\Theta^{-1}$**

⟨12⟩ *Coherent SI units* watt(s) per metre kelvin or kilogram metre(s) per second cubed kelvin (W m^{-1} K^{-1} = kg m s^{-3} K^{-1})

Term for unit	Symbol of unit		Value in coherent SI unit
	Recommended	Other	
kilowatt(s) per metre kelvin	kW m^{-1} K^{-1}		= 1000 W m^{-1} K^{-1}
calorie(s) per centimetre second degree Celsius		cal cm^{-1} s^{-1} °C^{-1}	= 418.68 W m^{-1} K^{-1}
kilocalorie(s) per metre hour degree Celsius		kcal m^{-1} h^{-1} °C^{-1}	= 1.163 W m^{-1} K^{-1}
watt(s) per metre kelvin	W m^{-1} K^{-1}		= 1 W m^{-1} K^{-1}
kilogram metre(s) per second cubed kelvin	kg m s^{-3} K^{-1}		= 1 kg m s^{-3} K^{-1}

9.77 Dimension L^{-2}M^{-1}T$^3\Theta$

9.77.1 ⟨2⟩ **thermal resistance** across an interface between subsystems 1 and 2
⟨5⟩ R
⟨6⟩ temperature difference across an interface, $\Delta_1^2 T$ divided by heat rate, $\Phi_{h,1,2}$
⟨7⟩ $R_{1,2} = \Delta_1^2 T / \Phi_{h,1,2}$
⟨9⟩ The term and symbol are sometimes used also for insulation coefficient (§9.74.1).

9.77U **Units of dimension L^{-2}M^{-1}T$^3\Theta$**

⟨12⟩ *Coherent SI units* kelvin(s) per watt or kelvin second cubed per kilogram square metre (K W^{-1} = K s^3 kg^{-1} m^{-2})

9.78 Dimension L^{2}MT$^{-3}\Theta^{-1}$

9.78.1 ⟨2⟩ **kelvic heat rate** across interface between subsystems 1 and 2
⟨3⟩ **thermal conductance; thermal conductivity; thermal conduction coefficient**
⟨5⟩ G
⟨7⟩ $G_{1,2} = 1/R_{1,2}$

9.78U **Units of dimension L^{2}MT$^{-3}\Theta^{-1}$**

⟨12⟩ *Coherent SI units* watt(s) per kelvin or kilogram square metre(s) per second cubed kelvin (W K^{-1} = kg m^2 s^{-3} K^{-1})

9.79 Dimension TJ

9.79.1 ⟨2⟩ **amount-of-light** of system 1
⟨3⟩ **luminous amount; quantity of light**
⟨5⟩ Q_v, Q
⟨6⟩ integral of light rate (or luminous flux) from, through or onto a system, $\Phi_{v,1}$, over time, t
⟨7⟩ $Q_{v,1} = \int \Phi_{v,1} \mathrm{d}t$

9.79U **Units of dimension TJ**

⟨12⟩ *Coherent SI units* lumen second(s) or candela steradian second(s) (lm s = cd sr s)

9.80 Dimension $L^{-2}TJ$

9.80.1 $\langle 2 \rangle$ **areic light** onto surface 1
$\langle 3 \rangle$ **luminous exposure; light exposure**
$\langle 5 \rangle$ H_v, H
$\langle 6 \rangle$ integral of areic light rate incident on a surface, $E_{v,1}$, over time, t
$\langle 7 \rangle$ $H_{v,1} = \int E_{v,1} dt$
$\langle 8 \rangle$ quantity of light
$\langle 9 \rangle$ The subscript v (for visual) is recommended to avoid confusion with radiometric or photon quantities.

9.80U **Units of dimension $L^{-2}TJ$**
$\langle 12 \rangle$ *Coherent SI units* lumen second(s) per square metre or lux second(s), or candela steradian second(s) per square metre ($lm\ s\ m^{-2} = lx\ s = cd\ sr\ s\ m^{-2}$)

9.81 Dimension J

9.81.1 $\langle 2 \rangle$ **light rate** of system 1
$\langle 3 \rangle$ **luminous flux**
$\langle 5 \rangle$ Φ_v, Φ
$\langle 6 \rangle$ luminous intensity (or steradic light rate) of a source over an element of solid angle, $I_{v,1}$, multiplied by that element of solid angle, Ω
$\langle 7 \rangle$ $d\Phi_{v,1} = I_{v,1}\ d\Omega_1$

9.81.2 $\langle 2 \rangle$ **luminous intensity**
$\langle 3 \rangle$ **steradic light rate**
$\langle 5 \rangle$ $I_{v,1}$, I
$\langle 9 \rangle$ (1) Luminous intensity is a base kind-of-quantity of SI.
$\langle 9 \rangle$ (2) For an element of solid angle, Ω, luminous intensity equals the amount-of-light from a source in that solid angle in a small time interval divided by that element and by the time difference:

$$I_{v,1} = d^2 Q_{v,1,\Omega}/(d\Omega\ dt)$$

9.81U **Units of dimension J**
$\langle 12 \rangle$ *Coherent SI units* lumen or candela steradian(s) ($lm = cd\ sr$)
$\langle 12 \rangle$ *SI base units* candela(s) (or lumen(s) per steradian or lumen(s)) ($cd = lm\ sr^{-1} = lm$)
$\langle 16 \rangle$ The candela is the luminous intensity, in a given direction, of a source that emits monochromatic radiation of frequency 540×10^{12} hertz and that has a radiant intensity in that direction of (1/683) watt per steradian (16th CGPM 1979, Resolution 3).

9.82 Dimension $L^{-2}M^{-1}T^3J$

9.82.1 $\langle 2 \rangle$ **luminous efficacy** of radiation λ
$\langle 5 \rangle$ K
$\langle 7 \rangle$ $K(\lambda) = Q_v(\lambda)/Q_e(\lambda)$
$\langle 9 \rangle$ Luminous intensity (or steradic light rate) of any source of light, I_v, can be calculated from the differential quotient of steradic energy rate to wavelength, $I_{e,\lambda}$, by the equation:

$$I_v = \int K(\lambda) I_{e\lambda}, d\lambda$$

The function $K(\lambda)$ is usually expressed as the product of the maximum luminous efficacy at frequency 540 THz (corresponding to wavelength about 555 nm) (in the definition of the candela, §9.81U) and a standardized factor for eye sensitivity with photopic vision, $V(\lambda)$, where $K(\lambda) = K_{max} V(\lambda)$.

Values of the standardized factor were published in the Procès-Verbaux of CIPM in 1972 (40th CIPM 1972 29, p. 145). These values do not apply for scotopic vision for which the maximum is at wavelength 507 nm. Estimates in the reverse direction are often necessary for calculation of areic energy rate of radiation in controlled biological systems, because commercial specifications of illumination sources are commonly provided on a photometric scale. Other biological effects of radiation, such as erythema and photosynthesis, have different functions of frequency or wavelength and cannot be predicted on a photometric scale.

9.82U   **Units of dimension $L^{-2}M^{-1}T^3J$**
$\langle 12 \rangle$ *Coherent SI unit* of luminous efficacy (cd m^{-2} kg^{-1} s^3)

9.83   Dimension N

9.83.1 $\langle 2 \rangle$ **amount-of-substance** of component B in system 1
$\langle 4 \rangle$ am.s.
$\langle 5 \rangle$ n_B, $n(B)$
$\langle 8 \rangle$ molar amount (deprecated by ISO/IEC 80000-1[5])
$\langle 9 \rangle$ (1) Amount-of-substance is a base kind-of-quantity of the International System of Units.
$\langle 9 \rangle$ (2) In developing a definition of the unit of amount-of-substance, the mole, the terms *elementary unit* and *formula unit* were variously used. The wider term **entity** is now preferred in order to avoid restricting use of the concept of amount-of-substance (§4.3.3).
$\langle 9 \rangle$ (3) Amount-of-substance and kinds-of-quantity derived from it should be used, in preference to mass or volume, whenever an entity can be specified (§4.3 and §7.1).
$\langle 9 \rangle$ (4) Amount-of-substance does not equal a number of entities but is proportional to a number of entities. The relationship is governed by the Avogadro constant (§9.84.1):

$$n_B = N_B/N_A$$

So 1 mole is proportional to about 0.602 214 129 $\times$ 10^{24} entities of specified type(s).
$\langle 9 \rangle$ (5) Amount-of-substance of a particular component is proportional to its mass, the proportionality factor being the reciprocal of **molar mass** (§9.92.1), which can be termed **massic substance** (§9.91.1, $\langle 9 \rangle$ (2)). The numeric values of molar mass in grams per mole equal values of '*atomic weight*' (or **relative molar mass** or *relative atomic mass*) (§8.4.5). Values are regularly revised by the IUPAC Commission on Atomic Weights and Isotopic Abundances. They are published in some of the IUPAC handbooks and in the journal *Pure and Applied Chemistry*.
$\langle 9 \rangle$ (6) Unless otherwise stated, amount-of-substance of a component here represents stoichiometric amount-of-substance, which may be distinguished $n_{o,B}$ (§7.2.3).
$\langle 9 \rangle$ (7) **Amount-of-substance of a chemical reaction** (example below), commonly termed **extent of reaction**, ξ_B, is defined with stoichiometric number positive for reaction products and negative for reactants, so that extent of reaction is always positive, in contrast to change in amount-of-substance of component B, n_B.
$\langle 9 \rangle$ (8) **Buffer capacity** has been defined only for H$^+$. Analogous kinds-of-quantity would apply for other components. For electroneutrality in measuring buffer capacity, H$^+$ is added together with an indifferent ionic component such as Cl$^-$ or is exchanged against a cation such as Na$^+$. Occasionally μ_B is taken as independent

variable instead of $n_{o,B}$; for instance, partial pressure or gas tension of CO_2 is taken instead of amount-of-substance of CO_2. This must be stated because it influences the value of buffer capacity.

$\langle 10 \rangle$ U—Nitrogen(N); am.s.(24 h) = 0.465 mol

$\langle 10 \rangle$ Syst—Osmotic entity; am.s.

$$n_{B,\ldots,N,1} = \Pi \ V/R \ T \quad \text{where } V \text{ is volume, } R \text{ is molar gas constant, and } T \text{ is}$$
thermodynamic temperature.

$\langle 10 \rangle$ Gas—Molecules; am.s.(pressure equivalent).

$$n_{A,\ldots,N,g} = p_g \ V_g/R \ T \quad \text{where } p_g \text{ is pressure exerted by a gas and } V_g \text{ is volume}$$
occupied by a gas.

$\langle 10 \rangle$ Syst—Reactant($|\nu_B^{-1}|$ B); am.s.

$$\xi_B = \nu_B^{-1} \ \Delta n_B \quad \text{where } \xi B \text{ is extent of reaction B.}$$

$\langle 10 \rangle$ Organism—Photons, incident($Q_e/n_\phi = 200$ kJ mol^{-1}); am.s.

$\langle 19 \rangle$ The definition of the mole is currently under discussion (CIPM, CCU), see §5.1.9.

9.83.1.1 $\langle 2 \rangle$ **entitic amount-of-substance** of component B in system 1
$\langle 4 \rangle$ entitic am.s.
$\langle 10 \rangle$ Trcs(B)—Serotonin; entitic am.s. = 0.012 amol

9.83U **Unit of dimension N**
$\langle 12 \rangle$ *SI base unit* mole(s) (mol)
$\langle 16 \rangle$ The mole is the amount-of-substance of a system which contains as many elementary entities as there are atoms in 0.012 kilogram of carbon-12 (14th CGPM 1971, Resolution 3).

| Term for unit | Symbol of unit | | Value in coherent SI unit |
	Recommended	Other	
mole(s)	mol	M, ᴍ	= 1 mol
gram-molecule(s)		g-mol	= 1 mol
gram-ion(s)		g-ion	= 1 mol
gram-atom(s)		g-at	= 1 mol
einstein(s)		E	= 1 mol
millimole(s)	mmol		= 0.001 mol
micromole(s)	μmol		= 10^{-6} mol
nanomole(s)	nmol		= 10^{-9} mol
picomole(s)	pmol		= 10^{-12} mol
femtomole(s)	fmol		= 10^{-15} mol
attomole(s)	amol		= 10^{-18} mol
zeptomole(s)	zmol		= 10^{-21} mol
yoctomole(s)	ymol		= 10^{-24} mol

$\langle 19 \rangle$ (1) In the definition of the mole, it is understood that 'atoms of carbon-12' refers to unbound atoms at rest and in their ground state.

$\langle 19 \rangle$ (2) Whenever the mole is used, the entities must be specified (§4.3.3), as in the examples of §9.83.1. This rider to the definition avoids the need for equivalents or gram-particles.

$\langle 19 \rangle$ (3) The unit was defined in relationship to the concept of 'atomic weight' in the sense of **relative atomic weight** (§8.4.5), for which the standard in chemistry used to be the natural mixture of isotopes of oxygen, given the value 16, whereas the standard in physics was oxygen-16. In 1959-1960, the discrepancy was ended by agreement between IUPAC and IUPAP, carbon-12 being taken as standard with the assigned value 12.

⟨19⟩ (4) The unrecognized symbols M or ᴍ were formerly used for both mole and mole per litre (§9.88U). They are still widely used in the second meaning, especially in biochemistry, but are not recognized by CGPM.

9.84   Dimension N^{-1}

9.84.1   ⟨2⟩ **Avogadro constant**
⟨5⟩ N_A, L
⟨6⟩ number of entities of any specified component, N_B, divided by the amount-of-substance of that component, n_B
⟨7⟩ $N_A = L = N_B/n_B \approx 0.602\ 214\ 129$ ymol^{-1}; $u_c = 0.000\ 000\ 027$ ymol^{-1} [ref. 15]
⟨9⟩ (1) The symbol N_A is used to honour A. Avogadro and L is in honour of J. Loschmidt.
⟨9⟩ (2) The assignment of a numerical value to the Avogadro constant is under discussion (CIPM, CCU), see §5.1.9.

9.84U   **Units of dimension N^{-1}**
⟨12⟩ *Coherent SI unit* (one) per mole or reciprocal mole(s) (mol^{-1})

9.85   Dimension LN^{-1}

9.85.1   ⟨2⟩ **integral of molar area decadic absorbance of component B over wavelength $\Delta \tilde{v}$** for system 1
⟨5⟩ $\varepsilon(\Delta\tilde{v})$
⟨7⟩ $\varepsilon(\Delta\tilde{v}) = \int_1^2 \varepsilon(\tilde{v})\mathrm{d}\tilde{v}$

9.85U   **Unit of dimension LN^{-1}**
⟨12⟩ *Coherent SI unit* metre(s) per mole (m mol^{-1})

9.86   Dimension L^2N^{-1}

9.86.1   ⟨2⟩ **molar area** of component B to surface 1 or interface of subsystems 1 and 2
⟨5⟩ $A_{m,B}$
⟨6⟩ area of a surface or interface, A, divided by the amount-of-substance of a component, n_B, in specified association to the surface or interface
⟨7⟩ $A_{m,B} = A_1/n_B$

9.86.2   ⟨2⟩ **molar area rotance** of optically active component B to linearly polarized light λ
⟨3⟩ **molar optical rotatory power**
⟨5⟩ $\alpha_{m,B}(\lambda)$, $\alpha_{n,B}(\lambda)$
⟨7⟩ $\alpha_{m,B}(\lambda) = \alpha_1(\lambda)/(c_B\ l)$

9.86.3   ⟨2⟩ **molar area absorbance** due to component B with radiation λ
⟨3⟩ **molar linear absorption coefficient; molar absorption coefficient; molar absorptivity**
⟨5⟩ $\varepsilon_B(\lambda)$
⟨6⟩ lineic absorbance of a radiation due to a component B divided by substance concentration of the component
⟨7⟩ $\varepsilon_B(\lambda) = a/c_B$ where a is the absorption coefficient and c_B is the amount-of-substance concentration.
⟨8⟩ absorption coefficient; absorptivity; extinction coefficient

⟨9⟩ The value depends on the scale of absorbance, commonly based on decadic logarithms in analytical chemistry and napierian logarithms in radiation physics.

⟨10⟩ Solution—Nicotinamide reduced; molar linear absorption coefficient(340 nm) = 630 m^2 mol^{-1}

9.86.3.1 ⟨2⟩ **molar area napierian absorbance** due to component B with radiation λ

⟨3⟩ **molar linear napierian absorption coefficient; molar napierian absorption coefficient; molar napierian absorptivity**

⟨5⟩ $\kappa_B(\lambda)$

⟨7⟩ $\kappa_B(\lambda) = \alpha_{l,B}(\lambda)/c_B$

⟨8⟩ absorption coefficient; absorptivity

9.86.3.2 ⟨2⟩ **molar area decadic absorbance** due to component B with radiation λ

⟨3⟩ **molar decadic absorption coefficient; molar linear decadic absorption coefficient; molar decadic absorptivity**

⟨5⟩ $\varepsilon_B(\lambda)$

⟨6⟩ lineic decadic absorbance of a radiation due to a component B, $a_B(\lambda)$, divided by its substance concentration, c_B

⟨7⟩ $\varepsilon_B(\lambda) = a_B(\lambda)/c_B$

$$= A_B(\lambda)/(c_B\ l)$$

⟨8⟩ absorption coefficient; absorptivity; † absorbance cross-section per mole

9.86U **Units of dimension L^{2}N^{-1}**

⟨12⟩ *Coherent SI unit of molar area* square metre(s) per mole (m^2 mol^{-1})

⟨12⟩ *Coherent SI units of molar area rotance* radian square metre(s) per mole or square metre(s) per mole (rad m^2 mol^{-1} = m^2 mol^{-1})

Term for unit	Symbol of unit		Value in coherent SI unit
	Recommended	Other	
litre(s) per millimole centimetre		L mmol^{-1} cm^{-1} or l mmol^{-1} cm^{-1}	= 100 m^2 mol^{-1}
square metre(s) per mole	m^2 mol^{-1}		= 1 m^2 mol^{-1}
square centimetre(s) per millimole		cm^2 mmol^{-1}	= 0.1 m^2 mol^{-1}
litre(s) per centimetre mole		L cm^{-1} mol^{-1} or l cm^{-1} mol^{-1}	= 0.1 m^2 mol^{-1}
square metre(s) per kilomole	m^2 kmol^{-1}		= 10^{-3} m^2 mol^{-1}
litre(s) per metre mole	L m^{-1} mol^{-1} or l m^{-1} mol^{-1}		= 10^{-3} m^2 mol^{-1}
square centimetre(s) per mole		cm^2 mol^{-1}	= 0.1 × 10^{-3} m^2 mol^{-1}

9.87 Dimension L^{-2}N

9.87.1 ⟨2⟩ **areic substance** of component B to surface 1 or interface of subsystems 1 and 2

⟨3⟩ **areic amount-of-substance**

⟨4⟩ areic am.s.

⟨5⟩ $n_{A,B}$

⟨6⟩ amount-of-substance of a component, n_B, in specified association to a surface or interface divided by the area of the surface or interface, A

⟨7⟩ $n_{A,B} = n_B/A_1$

9.87U   Units of dimension $L^{-2}N$

⟨12⟩ *Coherent SI unit* mole(s) per square metre (mol m^{-2})

9.88   Dimension $L^{-3}N$

9.88.1   ⟨2⟩ **substance concentration** of component B in system 1

⟨3⟩ **amount-of-substance concentration**; amount concentration; molar concentration

⟨4⟩ subst.c.

⟨5⟩ c_B, [B]

⟨6⟩ amount-of-substance of a component, n_B, divided by volume of the system, V_1

⟨7⟩ $c_B = n_B/V_1$

⟨8⟩ concentration; molecular concentration

⟨9⟩ (1) Substance concentration can be calculated from mass concentration, ρ_B, by means of molar mass, M_B:

$$c_B = \rho_B\, M_B$$

It also equals the product of molality, b_B, and mass concentration of solvent, ρ_A:

$$c_B = b_B\, \rho_A$$

⟨8⟩ *molarity* (discouraged due to the risk of confusion with molality)

⟨9⟩ (2) Substance concentration depends on temperature and, for gaseous systems, on the pressure of the system. Expression of results as substance content or molality avoids that variability.

⟨9⟩ (3) The entity of component should be stated (§4.3.3).

⟨9⟩ (4) The concept of substance concentration has supplanted the ambiguous concept of equivalent concentration. Equivalence of substance concentrations of different reacting components depends on choice of ionic charge entity or stoichiometric entity.

⟨9⟩ (5) For the continuing history of terms of this kind-of-quantity, see §5.4.14.

⟨9⟩ (6) The concepts of amount-of-substance and substance concentration are crucial to clinical laboratory sciences (biological chemistry, biological pharmacology, clinical toxicology, ...) as described in §7.1.

⟨9⟩ (7) Substance concentration of a component B usually means the stoichiometric substance concentration (§7.2.3) and not active substance concentration (§9.88.3) of free component.

⟨9⟩ (8) The term *ionic concentration* usually means $c(B^z)$ and not $c(z_B^{-1}B^z)$.

⟨10⟩ For Ca^{2+} in blood plasma:

$c(Ca^{2+}) = a_b(Ca^{2+})\ \gamma(Ca^{2+})^{-1}\varrho(H_2O)\ \tilde{b}_\theta$ where $\tilde{b}_e = 1$ mol kg^{-1}. The appropriate constant:

$k(Ca^{2+}) = \gamma(Ca^{2+})^{-1}\ \rho(H_2O)\ \tilde{b}_e$ where the mean value of $\gamma(Ca^{2+})^{-1}\ \rho(H_2O)$ for normal blood plasma should be taken.

⟨10⟩ B(Person)—Haemoglobin(Fe); subst.c. $= 9.1$ mmol L^{-1}

⟨10⟩ B—Dioxygen(total); subst.c.

$c(tO_2) = c(HbO_2) + \alpha(O_2) \times p(O_2)$

⟨10⟩ P(B)—Carbon dioxide; subst.c. $= 7.1$ mmol L^{-1}

⟨10⟩ P(B)—Hydrogen carbonate ion; subst.c.(Blood; $pCO_2 = 5.3$ kPa; 37 °C) $= 24.2$ mmol L^{-1}

⟨10⟩ Syst—Redox component B; subst.c. $= ?$ mmol L^{-1}

⟨10⟩ Lactate dehydrogenase measuring system—L-Lactate; subst.c. (IFCC 2002) = 50 mmol L^{-1}

⟨9⟩ (9) according to IFCC primary reference procedure[16]

⟨10⟩ Air(ambient)—Antimony(O + III + V); subst.c. = 0.1 nmol m^{-3}

⟨9⟩ (10) The denominator m^3 is used for concentrations in environmental air instead of litre.

9.88.1.1 ⟨2⟩ **substance concentration of ionic charge entity $|z_B{}^{-1}|$ B**

⟨5⟩ $c(|z_B{}^{-1}|$ B)

⟨8⟩ normality; ionic concentration

⟨10⟩ Syst—Ionic-charge entity $|z_B{}^{-1}|$ B); subst.c. = ? mmol L^{-1}

9.88.1.2 ⟨2⟩ **ionic strength of entity in terms of substance concentration**

⟨3⟩ **concentrational ionic strength**

⟨5⟩ I_c

⟨7⟩ $I_c = 0.5 \sum_{B}^{N} c(|zB^2|$ B)

⟨8⟩ ionic strength; I_B

⟨10⟩ Syst—Cationic charge entities; subst.c. = ? mmol L^{-1}

⟨9⟩ See also §7.1.6, §9.83.1, Note 9, and §9.91.2.1.

9.88.2 ⟨2⟩ **differential quotient of stoichiometric substance concentration of hydrogen ion to pH**

⟨3⟩ **volumic buffer capacity**

⟨5⟩ β

⟨7⟩ $\beta = B/V_1$ where B is amount-of-substance of volumic buffer capacity (§9.83.1).
$$= [dc_{\ominus}(H^+)/d\lg a_b(H^+)]_V = dc(\text{base})/d\text{pH}$$

⟨8⟩ buffer capacity; buffer value

⟨10⟩ EcF—Base excess; subst.c.(actual – norm) = + 1.8 mmol L^{-1}

9.88.3 ⟨2⟩ **active substance concentration** of solute B in solution 1

⟨5⟩ $\tilde{c}_B$

⟨7⟩ $\tilde{c}_B = y_B\, c_B$ where y_B is substance-concentrational activity factor (§8.10.3.3).

⟨8⟩ concentrational activity

⟨9⟩ (1) $\lim_{x(A) \to 1} \tilde{c}_B = c_B$

⟨9⟩ (2) $\tilde{c}_B/\tilde{b}_B = \rho_{A^*}$ where the star indicates 'pure'.

⟨9⟩ (3) Results obtained with ion-selective electrodes may be reported as 'substance concentrations' of the ion (§7.2.4 and §7.2.10). They should not be expressed analogously to pH (§8.9.4). The 'substance concentration' of the ion is calculated from the product of the molal activity and an appropriate constant depending on the species of ion and the type of biological fluid under investigation. Ion-selective electrodes may, alternatively, be calibrated in a similar way to pH electrodes, *i.e.* on the basis of molal activity.

9.88.3.1 ⟨2⟩ **osmolarity; osmotic concentration** of solution 1

⟨5⟩ $\hat{c}$

⟨6⟩ (1) product of the osmolality of a solution, $\hat{b}$, and the volumic mass of the solvent (commonly water), ρ_{A^*} where the star indicates 'pure'

⟨7⟩ $\hat{c} = \hat{b}\rho_{A^*}$

⟨6⟩ (2) negative logarithm of the relative chemical activity of a solvent, a_A, divided by the molar volume of the solvent, $V_{m,A*}$ where the star indicates 'pure'.

⟨7⟩ $\hat{c} = (-\ln a_A)/V_{m,A*}$

⟨8⟩ osM, osmol, osmol L^{-1}

⟨9⟩ The SI unit for osmolarity used in clinical laboratory sciences is mol L^{-1}. Osm kg^{-1} and osmol kg^{-1} cannot be used for osmolarity.

9.88.4   ⟨2⟩ **substance-concentrational equilibrium coefficient** of reaction B

⟨3⟩ **concentrational equilibrium product**

⟨5⟩ $K_{c,B}$

⟨6⟩ product of the substance concentrations of the reactants in a reaction to the power of their stoichiometric numbers

⟨8⟩ equilibrium constant

⟨9⟩ (1) The coefficient is of variable dimension $(L^{-3}N)^n$, where $n = \sum\limits_{B}^{N} \nu B$.

⟨9⟩ (2) Other equilibrium coefficients are definable in terms of chemical activity (§8.10.1), substance fraction, partial pressure (§9.41.1.3) or molality (§9.91.2).

9.88.4.1   ⟨2⟩ **Michaelis–Menten coefficient**

⟨3⟩ Michaelis–Menten constant; Michaelis constant

⟨5⟩ K_M

⟨6⟩ equivalent to the substrate concentration at which the reaction takes place at one half its maximum rate

⟨7⟩ $K_M = c_B c_E/c_{BE}$

for the formation of a binary substrate–enzyme complex $B + E \rightarrow BE$

⟨9⟩ The same term applies to ternary or more complex reactions with coefficients, for instance, of dimension $L^{-6}N^2$.

9.88U   **Units of dimension $L^{-3}N$**

⟨12⟩ *Coherent SI unit* mole(s) per cubic metre (mol m^{-3})

Term for unit	Symbol of unit Recommended	Other	Value in coherent SI unit
kilomole(s) per cubic metre	kmol m^{-3}		$= 1000$ mol m^{-3}
mole(s) per litre	mol L^{-1}, mol l^{-1}		$= 1000$ mol m^{-3}
molar		M, M	$= 1000$ mol m^{-3}
mole(s) per cubic metre	mol m^{-3}		$= 1$ mol m^{-3}
millimole(s) per litre	mmol L^{-1}, mmol l^{-1}		$= 1$ mol m^{-3}
millimolar		mM, mM	$= 1$ mol m^{-3}
millimole(s) per cubic metre	mmol m^{-3}		$= 0.001$ mol m^{-3}
micromole(s) per litre	μmol L^{-1}, μmol l^{-1}		$= 0.001$ mol m^{-3}
micromolar		μM, μM	$= 0.001$ mol m^{-3}
micromole(s) per cubic metre	μmol m^{-3}		$= 10^{-6}$ mol m^{-3}
nanomole(s) per litre	nmol L^{-1}, nmol l^{-1}		$= 10^{-6}$ mol m^{-3}

⟨19⟩ (1) For considerations in the choice of litre or cubic metre for expression of concentrations see §5.10.2.

⟨19⟩ (2) The unit molar (M, M) is accepted by the IUBMB and by ISO/IEC 80000-1,[5] but is not recognized by the CGPM.

9.89 Dimension L^3N^{-1}

9.89.1 ⟨2⟩ **partial molar volume** of component B in gaseous system 1
⟨4⟩ molar vol.
⟨5⟩ $V_{m,B}$
⟨6⟩ change in volume of a system due to a component, V_1, divided by change in amount-of-substance of molecules, n_B
⟨7⟩ $V_{m,B} = (\partial V_1 / \partial n_B)$
⟨8⟩ V_B
⟨9⟩ (1) The partial molar volume of an ideal gas can be calculated from
$$V_{m,id}\, R \times T/p$$
At $T = 273.15$ K and $p = 101\ 325$ Pa:
$$V_{m,id} = 22\ 413.968 \times 10^{-6}\ \text{m}^3\ \text{mol}^{-1};\ u_c = 0.020 \times 10^{-6}\ \text{m}^3\ \text{mol}^{-1}$$
⟨9⟩ (2) The partial volume of a system, V_1, equals the sum of the products of molar volume and amount-of-substance, n_B, of each component:

$$V_1 = \sum_{A}^{N} (V_{m,B}\, n_B)$$

9.89.2 ⟨2⟩ **integral of molar area napierian absorbance over wavelength** for radiation $\Delta\lambda$
⟨5⟩ $A(\Delta\lambda)$
⟨7⟩ $A(\Delta\lambda) = \int_1^2 \kappa(\lambda)\mathrm{d}\lambda$

9.89U **Unit of dimension L^3N^{-1}**
⟨12⟩ *Coherent SI unit* cubic metre(s) per mole ($\text{m}^3\ \text{mol}^{-1}$)

9.90 Dimension $L^{-4}N$

9.90.1 ⟨2⟩ **substance concentration gradient** of component B in direction x
⟨4⟩ subst.c. gradient
⟨5⟩ $\mathbf{grad}_x\, c_B$, ∇c_B
⟨6⟩ change in substance concentration of a component, c_B, in a small distance x divided by that distance
⟨7⟩ $\nabla c_B = \mathrm{d}c_B / \partial x, \partial y, \partial z$

9.90U **Unit of dimension $L^{-4}N$**
⟨12⟩ *Coherent SI unit* mole(s) per metre to the power four ($\text{mol}\ \text{m}^{-4}$)

9.91 Dimension $M^{-1}N$

9.91.1 ⟨2⟩ **substance content** of component B in system 1
⟨4⟩ subst.cont.
⟨5⟩ ν_B
⟨6⟩ amount-of-substance of a component, n_B, divided by mass of the system, m_1
⟨7⟩ $\nu_B = n_B / m_1$
⟨8⟩ specific mole content; content
⟨9⟩ (1) Such a quantity is independent of temperature and pressure. Therefore it has some advantages over substance concentration (§9.88.1). It is more informative than mass fraction (§8.4.1).
⟨9⟩ (2) For a pure substance, substance content equals **massic substance**, which is the reciprocal of molar mass (§7.1.4, Table 7.2).
⟨9⟩ (3) Substance content is often confused with molality (§9.91.2).
⟨10⟩ Faes—Calcium; subst.cont. $= 15$ mmol kg^{-1}
⟨10⟩ Hair—Arsenic; subst.cont. $= 5.1$ μmol kg^{-1}

9.91.2   ⟨2⟩ **molality** of solute component B in solution 1
⟨4⟩ molal.
⟨5⟩ b_B, m_B
⟨6⟩ amount-of-substance of a solute, n_B, divided by mass of solvent, m_A
⟨7⟩ $b_B = n_B/m_A$
⟨9⟩ (1) Molality differs in type of definition from substance content and substance concentration, in which the mass and volume, respectively, in the denominator refer to the system or the mixture, whereas for molality it refers only to solvent.
⟨9⟩ (2) Such a quantity is independent of the pressure and temperature of the solution. This property sometimes weighs in favour of using molality instead of substance concentration, *e.g.* when stating the composition of solutions employed for freezing-point determinations. Substance content (§9.91.1) has some of the same advantages.
⟨9⟩ (3) The entity of solute should be stated, if doubt may arise.
⟨9⟩ (4) Provided that all chemical components involved in the mixture are known, molality can be converted to substance fraction, x_B, which is of central importance in physical chemistry:

$$b_B = x_B/(x_A\, M_A) \quad \text{where } M_A \text{ is molar mass of solvent.}$$

There is a relationship between molality, substance concentration and mass concentration of solvent:

$$b_B = c_B/\rho_A \quad \text{where } \rho_A \text{ is mass concentration of solvent.}$$

⟨9⟩ (5) Molality is sometimes confused with *molarity*, an old term for substance concentration (§9.88.1).
⟨9⟩ (6) Because of the ambiguity of the symbol m for molality and mass, the symbol b is used for molality in the definitions of this Compendium, as favoured by ISO/IEC 80000-5.[14] In clinical chemistry, the symbol b is also employed for catalytic concentration.
⟨10⟩ P—Solute; molal.(proc.) $= 314$ mmol kg^{-1}

9.91.2.1   ⟨2⟩ **ionic strength (molality basis)** in solution 1
⟨3⟩ average ionic molality; **molal ionic strength**
⟨5⟩ I_b, I_m
⟨7⟩ $I_b = 0.5\ \Sigma_B(z_B{}^2 \cdot b_B)$
⟨8⟩ ionic strength; I; $m_\pm$
⟨9⟩ For explanation see also §7.1.6.

9.91.2.2   ⟨2⟩ **active molality** of solute B in solution 1
⟨5⟩ $\tilde{b}_B$, $\tilde{m}_B$
⟨7⟩ $\tilde{b}_B = \gamma_B\, b_{B,1}$ where γ_B is molal activity factor (§8.10.3.2).
⟨8⟩ molal activity
⟨9⟩ $\lim_{x(A) \to 1} \tilde{b}_B = b_B$

9.91.2.3   ⟨2⟩ **osmolality** of solution 1
⟨5⟩ $\hat{b}$, $\hat{m}$
⟨6⟩ negative napierian logarithm of the substance-fractional relative activity of solvent (usually water), a_A, divided by its molar mass, M_A
⟨7⟩ $\hat{b} = -(\ln a_A)/M_A$
⟨8⟩ osM, osmol, osmol/kg
⟨9⟩ (1) $\tilde{p}_A = \tilde{p}_{A^*}\, \exp(-\hat{b}M_A)$ where the star indicates 'pure'.
⟨9⟩ (2) The coherent SI unit for osmolality is mol kg^{-1}. Osmol L^{-1} and osmol L^{-1} cannot be used for osmolality.

9.91.3 ⟨2⟩ **molal equilibrium coefficient**
⟨3⟩ **equilibrium constant (molality basis); molal equilibrium product**
⟨5⟩ K_b, K_m

⟨7⟩ $K_b = \prod\limits_{B}^{N} b_B{}^n$ where $n = \sum\limits_{B}^{N} \nu B$

⟨9⟩ (1) This group of kinds-of-quantity has the variable dimension $(M^{-1}N)^n$.
⟨9⟩ (2) Other equilibrium coefficients exist defined in terms, for instance, of substance fraction, chemical activity (§8.10.1–2), pressure or substance concentration (§9.88.1).

9.91U **Units of dimension $M^{-1}N$**
⟨12⟩ *Coherent SI unit* mole(s) per kilogram (mol kg^{-1})

Term for unit	Symbol of unit		Value in coherent SI unit
	Recommended	Other	
mole(s) per 100 grams		mol/100 g	$= 10$ mol kg^{-1}
mole(s) per kilogram	mol kg^{-1}		$= 1$ mol kg^{-1}
millimole(s) per gram		mmol g^{-1}	$= 1$ mol kg^{-1}
micromole(s) per milligram		µmol mg^{-1}	$= 1$ mol kg^{-1}
millimole(s) per kilogram	mmol kg^{-1}		$= 0.001$ mol kg^{-1}
micromole(s) per gram		µmol g^{-1}	$= 0.001$ mol kg^{-1}
micromole(s) per kilogram	µmol kg^{-1}		$= 10^{-6}$ mol kg^{-1}
nanomole(s) per gram		nmol g^{-1}	$= 10^{-6}$ mol kg^{-1}

⟨19⟩ (1) A solution of molality 1 mol kg^{-1} is sometimes termed a 1-molal solution. The unit is not recognized by CGPM and is not recommended and the symbol m must not be used for mol kg^{-1} in combination with other units.
⟨19⟩ (2) The kilogram is preferred as denominator in units (§5.9.8).

9.92 Dimension MN^{-1}

9.92.1 ⟨2⟩ **molar mass** of component B
⟨5⟩ M_B
⟨6⟩ mass of a component, m_B, divided by its amount-of-substance, n_B
⟨7⟩ $M_B = m_B/n_B$
⟨9⟩ see §9.92U ⟨19⟩ (2).
⟨10⟩ Human—Albumin; molar m. $= 66\ 437$ g mol^{-1}
⟨19⟩ The words "average molar mass" are frequently used for molar mass of macromolecules (§4.3.7).

9.92U **Unit of dimension MN^{-1}**
⟨12⟩ *Coherent SI unit* kilogram(s) per mole (kg mol^{-1})
⟨19⟩ (1) Published values of atomic weight and molecular weight (or relative molar mass, §8.6.5), $M_{r,B}$, has the same numerical values as molar mass in grams per mole (g $mol^{-1} = 10^{-3}$ kg mol^{-1}).
⟨19⟩ (2) A molecular mass m_B/N_B (§9.8.2.2) or particle mass expressed in daltons, Da, or atomic mass units, u, has the same numerical value as a molar mass expressed in g mol^{-1} (§5.6.5).

9.93   Dimension $T^{-1}N^{-1}$

9.93.1 $\langle 2 \rangle$ **molar absorbed dose** of component B in system 1
$\langle 5 \rangle$ $A_{m,B}$
$\langle 7 \rangle$ $A_{m,B} = A_B/n_B$
$\langle 8 \rangle$ molar radioactivity, specific activity

9.93U   **Units of dimension $T^{-1}N^{-1}$**
$\langle 12 \rangle$ *Coherent SI units* becquerel(s) per mole or (one) per second mole
$(\text{Bq mol}^{-1} = \text{s}^{-1} \text{ mol}^{-1})$

9.94   Dimension $T^{-1}N$

9.94.1 $\langle 2 \rangle$ **substance rate** of component B in system 1 or across interface between subsystems 1 and 2
$\langle 3 \rangle$ **substance flow rate**
$\langle 4 \rangle$ subst.rate
$\langle 5 \rangle$ $\dot{n}_B$, $F_{n,B}$, $\Phi_{n,B}$
$\langle 6 \rangle$ amount-of-substance of a component, n_B, changed in or moved to or from a system in a small time interval, t, divided by the time difference
$\langle 7 \rangle$ $\dot{n}_B = dn_B/dt$ in general
$\qquad\quad = \Delta n_B/\Delta t$ for an average substance rate or for a constant process
$\langle 8 \rangle$ mole rate
$\langle 10 \rangle$ Diet(Person)—Manganese(II); subst.rate(recommended) = 42 µmol d^{-1}
$\langle 10 \rangle$ Pt(U)—Glucose excretion; subst.rate(3 d) = 0.24 mmol d^{-1}
$\langle 10 \rangle$ Pt—Retinol absorption; subst.rate(proc.) = 3.1 µmol d^{-1}
$\langle 10 \rangle$ Pt(U)—Lead; subst.rate(proc.) = 22 nmol d^{-1}

9.94.2 $\langle 2 \rangle$ **substance rate of reaction** ν^{-1} of component B in system 1
$\langle 5 \rangle$ $\dot{n}$ $(\nu^{-1}B)$, $\dot{\xi}_B$
$\langle 6 \rangle$ change in amount-of-substance of a reactant or reaction product $\nu^{-1}B$ (or extent of reaction), ξ_B, in a small time interval, t, divided by the time difference
$\langle 7 \rangle$ $\dot{\xi}_B = d\xi_B/dt$
$\qquad\quad = \nu_B^{-1} dn_B/dt$ where ν_B is stoichiometric number of a reactant or product B.
$\langle 8 \rangle$ rate of reaction; reaction rate; rate of conversion; ν_B

9.94.2.1 $\langle 2 \rangle$ **catalysed substance rate of reaction** of product B in measuring system 1
$\langle 3 \rangle$ **catalysed rate of reaction**
$\langle 5 \rangle$ $\dot{n}$ $(\nu^{-1}B,\text{cat})$, $\dot{\xi}_{B,\text{cat}}$
$\langle 6 \rangle$ substance rate of a reactant or reaction product in the presence of catalyst, $\dot{\xi}_1$, minus that under identical conditions apart from the absence of catalyst, $\dot{\xi}_0$
$\langle 7 \rangle$ $\dot{\xi}_{B,\text{cat}} = \dot{\xi}_1 - \dot{\xi}_0$

9.94.3 $\langle 2 \rangle$ **catalytic activity** of catalytic component E in biological system 1
$\langle 3 \rangle$ **enzyme activity; enzym(at)ic activity; catalytic amount**
$\langle 4 \rangle$ cat.act.
$\langle 5 \rangle$ z_E
$\langle 6 \rangle$ (1) property of a biological component measured by the catalysed substance rate of conversion of a specified chemical reaction, in a specified measuring system
$\langle 6 \rangle$ (2) increase in substance rate of conversion of reactant or reaction product $\nu^{-1}B$ in a reaction catalysed by an enzyme in a specified measuring system
$\langle 8 \rangle$ amount of enzyme; amount-of-catalyst; K

⟨9⟩ Although the concept applies to catalysts in general, it was developed specifically for the needs of biochemists and clinical chemists in investigation of enzymes, *i.e.* organic catalysts. See further §7.6.
⟨10⟩ Ercs(B)—Glutathione peroxidase; cat.act.(37 °C; proc.) = ? akat

9.94U   Units of dimension $T^{-1}N$

⟨12⟩ *Coherent SI unit of substance rate* mole(s) per second (mol s^{-1})
⟨12⟩ *Coherent SI units of catalytic activity* mole(s) per second or katal(s) (kat = mol s^{-1})

Term for unit	Symbol of unit Recommended	Other	Value in coherent SI unit
mole(s) per second	mol s^{-1}		= 1 mol s^{-1}
katal(s)	kat	cat	= 1 kat = 1 mol s^{-1}
mole(s) per minute		mol min^{-1}	≈ 16.666 67 × 10^{-3} mol s^{-1}
millimole(s) per second	mmol s^{-1}		= 0.001 mol s^{-1}
millikatal(s)	mkat		= 0.001 kat
mole(s) per hour		mol h^{-1}	≈ 0.277 8 × 10^{-3} mol s^{-1}
mole(s) per day	mol d^{-1}		≈ 11.574 07 × 10^{-6} mol s^{-1}
micromole(s) per second	μmol s^{-1}		= 10^{-6} mol s^{-1}
microkatal(s)	μkat		= 10^{-6} kat
mole(s) per year	mol a^{-1}	mol year^{-1}	≈ 31.688 76 × 10^{-9} mol s^{-1}
enzyme unit(s)		U	≈ 16.666 67 × 10^{-9} kat
micromole(s) per minute		μmol min^{-1}	≈ 16.666 67 × 10^{-9} kat
nanomole(s) per second	nmol s^{-1}		= 10^{-9} mol s^{-1}
nanokatal(s)	nkat		= 10^{-9} kat

⟨19⟩ On coherence in units of time see §9.17U, ⟨19⟩ (2).

9.95   Dimension $L^{-2}T^{-1}N$

9.95.1   ⟨2⟩ **areic substance rate** of component B across interface between subsystem 1 and 2
⟨3⟩ substance flux density
⟨4⟩ areic subst.rate
⟨5⟩ $J_{n,B}$, $\varphi_{n,B}$
⟨6⟩ substance rate of a component across an interface, $\dot{n}_{B,1,2}$, divided by the area of the interface, $A_{1,2}$
⟨7⟩ $J_{n,B,1,2} = \dot{n}_{B,1,2}/A_{1,2}$
⟨8⟩ substance flux
⟨10⟩ Surface—Photon, incident; areic subst.rate = ? mol m^{-2} s^{-1}
⟨9⟩ Synonyms: *photon flux density*; *photon flux*; *quantum flux density*; *quantum flux*; $\varphi_{n,\Phi}$

9.95U   Units of dimension $L^{-2}T^{-1}N$

⟨12⟩ *Coherent SI unit* mole(s) per square metre second (mol m^{-2} s^{-1})

Term for unit	Symbol of unit Recommended	Other	Value in coherent SI unit
mole(s) per square centimetre second		mol cm^{-2} s^{-1}	= 10 × 10^3 mol m^{-2} s^1
kilomole(s) per square metre second	kmol m^{-2} s^{-1}		= 1000 mol m^{-2} s^{-1}
mole(s) per square metre second	mol m^{-2} s^{-1}		= 1 mol m^{-2} s^{-1}

9.96 Dimension $L^{-3}TN$

9.96.1 ⟨2⟩ **time integral of substance concentration** of component B in system 1
⟨7⟩ $\int c_B dt$
⟨8⟩ concentration-time product
⟨9⟩ This kind of integral can be used as a measure of exposure of organisms to biologically active components in the environment, for instance occupational exposure to atmospheric carcinogens that may be inhaled or contaminate the skin.

9.96U **Unit of dimension $L^{-3}TN$**
⟨12⟩ *Coherent SI unit* mole second(s) per cubic metre (mol s m^{-3})

9.97 Dimension $L^{-3}T^{-1}N$

9.97.1 ⟨2⟩ **substance concentration rate of reactant** $\nu^{-1}B$ in system 1
⟨3⟩ rate of reaction based on amount concentration
⟨5⟩ ν_B
⟨6⟩ substance rate of a reaction component, $\dot{n}(\nu'^{-1}B)$, divided by volume of the system, V_1
⟨7⟩ $\nu_B = \dot{n}(\nu'^{-1}B)/V_1$ where ν' is stoichiometric number of a reactant B
    $= \dot{\xi}_B/V_1$
⟨8⟩ rate of reaction

9.97.2 ⟨2⟩ **catalytic-activity concentration** of catalytic component E in biological system 1
⟨3⟩ **catalytic concentration**
⟨4⟩ cat.c.
⟨5⟩ b_E, κ_E, e_E
⟨6⟩ catalytic activity of a component, z_E, divided by volume of the system, V_1 (*i.e.* the original system containing the enzyme, not the measuring system)
⟨8⟩ enzyme concentration; concentration of enzyme activity; catalyst concentration; enzyme activity
⟨9⟩ (1) Besides the effect of temperature on catalytic activity, temperature affects the volume and therefore the catalytic-activity concentration.
⟨9⟩ (2) The symbol b is used also for molality.
⟨9⟩ (3) The measurement procedure is a necessary part of the specification.
⟨10⟩ P-Aspartate transaminase(EC 2.6.1.1); cat.c.(IFCC 2002) $= 5.1 \times 10^{-6}$ kat L^{-1}

9.97U **Units of dimension $L^{-3}T^{-1}N$**
⟨12⟩ *Coherent SI unit of substance concentration rate of reactant* mole(s) per cubic metre second (mol m^{-3} s^{-1})
⟨12⟩ *Coherent SI units of catalytic-activity concentration* katal(s) per cubic metre or mole(s) per cubic metre second (kat m^{-3} = mol m^{-3} s^{-1})

Term for unit	Symbol of unit		Value in coherent SI unit
	Recommended	Other	
kilomole(s) per cubic metre second	kmol m^{-3} s^{-1}		= 1000 mol m^{-3} s^{-1}
mole(s) per litre second	mol L^{-1} s^{-1}		= 1000 mol m^{-3} s^{-1}
katal(s) per litre	kat L^{-1}, kat l^{-1}		= 1000 kat m^{-3}
mole(s) per cubic metre second	mol m^{-3} s^{-1}		= 1 mol m^{-3} s^{-1}

(*Continued*)

Term for unit	Symbol of unit		Value in coherent SI unit
	Recommended	Other	
millimole(s) per litre second	mmol L^{-1} s^{-1} or mmol l^{-1} s^{-1}		$=1$ mol m^{-3} s^{-1}
katal(s) per cubic metre	kat m^{-3}		$=1$ kat m^{-3}
millikatal(s) per litre	mkat L^{-1}, mkat l^{-1}		$=1$ kat m^{-3}
micromole(s) per millilitre minute		μmol mL^{-1} min^{-1} or μmol ml^{-1} min^{-1}	$\approx 16.67 \times 10^{-3}$ mol m^{-3} s^{-1}
enzyme unit(s) per millilitre		U mL^{-1}, U ml^{-1}	$\approx 16.67 \times 10^{-3}$ kat m^{-3}
micromole(s) per litre second	μmol L^{-1} s^{-1} or μmol l^{-1} s^{-1}		$=0.001$ mol m^{-3} s^{-1}
millikatal(s) per cubic metre	mkat m^{-3}		$=0.001$ kat m^{-3}
microkatal(s) per litre	μkat L^{-1}, μkat l^{-1}		$=0.001$ kat m^{-3}
micromole(s) per litre minute		μmol L^{-1} min^{-1} or μmol l^{-1} min^{-1}	$\approx 16.67 \times 10^{-6}$ mol m^{-3} s^{-1}
enzyme unit(s) per litre		U L^{-1}, U l^{-1}	$\approx 16.67 \times 10^{-6}$ kat m^{-3}
nanomole(s) per litre second	nmol L^{-1} s^{-1} or nmol l^{-1} s^{-1}		$=10^{-6}$ mol m^{-3} s^{-1}
microkatal(s) per cubic metre	μkat m^{-3}		$=10^{-6}$ kat m^{-3}
nanokatal(s) per litre	nkat L^{-1}, nkat l^{-1}		$=10^{-6}$ kat m^{-3}

⟨19⟩ IUPAC, IFCC, ICSH and WASP[1,3] recommend the denominator unit litre for concentrations (§5.10.2) and the special SI term katal (Table 5.5), thus here katal per litre for the material with enzyme component rather than mole per cubic metre second or mole per litre second, which are used for the substance concentration rate of reactant in the measuring system.

9.98 Dimension M^{-1}T^{-1}N

9.98.1 ⟨2⟩ **substance content rate** of component B in system 1
⟨4⟩ subst.cont.rate
⟨5⟩ ν_B
⟨6⟩ change in substance content of a component, ν_B, in a small time interval, t, divided by the time difference
⟨7⟩ $\nu_B = d\nu_B/dt$
$\quad = d(n_B/m_1)/dt$
⟨10⟩ Skin FibroblProt—Copper(II) uptake; subst.cont.rate(6.5 h) $= 6.2$ nmol kg^{-1} s^{-1}

9.98.2 ⟨2⟩ **catalytic-activity content** of catalytic component E in system 1
⟨3⟩ **catalytic content**
⟨4⟩ cat.cont.
⟨6⟩ catalytic activity of a component, z_E, divided by the mass of the system, m_1
⟨7⟩ z_E/m_1
⟨10⟩ Faes—Chymotrypsin; cat.cont.(proc.) $= 0.3$ mkat kg^{-1}

9.98U   **Units of dimension $M^{-1}T^{-1}N$**
⟨12⟩ *Coherent SI unit of substance content rate* mole(s) per kilogram second $(\text{mol kg}^{-1}\,\text{s}^{-1})$
⟨12⟩ *Coherent SI units of catalytic-activity content* katal(s) per kilogram or mole(s) per kilogram second $(\text{kat kg}^{-1} = \text{mol kg}^{-1}\,\text{s}^{-1})$

9.99   Dimension $T^{-2}N$

9.99.1   ⟨2⟩ **catalytic-activity rate** of catalytic component E in system 1
⟨4⟩ cat.rate
⟨5⟩ $\dot{z}_E$
⟨6⟩ change in catalytic activity of a component, z_E, in a small time interval, t, divided by the time difference
⟨7⟩ $\dot{z}_E = dz_E/dt$
⟨10⟩ Pancreas—α-Amylase production; cat.rate(IFCC 2006) $= 18\ \mu\text{kat s}^{-1}$

9.99U   **Units of dimension $T^{-2}N$**
⟨12⟩ *Coherent SI units* katal(s) per second or mole(s) per cubic metre second squared $(\text{kat s}^{-1} = \text{mol s}^{-2})$

9.100   Dimension $L^{-3}T^{-2}N$

9.100.1   ⟨2⟩ **catalytic-activity concentration rate** of catalytic component E in system 1
⟨4⟩ cat.c.rate
⟨7⟩ $d(z_E/V_1)/dt$

9.100U   **Units of dimension $L^{-3}T^{-2}N$**
⟨12⟩ *Coherent SI units* katal(s) per cubic metre second or mole(s) per cubic metre second squared $(\text{kat m}^{-3}\,\text{s}^{-1} = \text{mol m}^{-3}\,\text{s}^{-2})$

9.101   Dimension $M^{-1}T^{-2}N$

9.101.1   ⟨2⟩ **catalytic-activity content rate** of catalytic component E in system 1
⟨4⟩ cat.cont.rate
⟨7⟩ $d(z_E/m_1)/dt$

9.101U   **Units of dimension $M^{-1}T^{-2}N$**
⟨12⟩ *Coherent SI units* katal(s) per kilogram second or mole(s) per kilogram second squared $(\text{kat kg}^{-1}\,\text{s}^{-1} = \text{mol kg}^{-1}\,\text{s}^{-2})$

9.102   Dimension $L^{-2}M^{-1}T^{2}N$

9.102.1   ⟨2⟩ **substance-concentrational solubility coefficient** of gaseous solute B in solution 1
⟨7⟩ $\alpha_{c,B} = c_B/\tilde{p}_B$
$= \alpha_{c,B,\infty}/y_B$ where $\tilde{p}_B$ is active partial pressure (§9.41.3) and y_B is substance-concentrational activity factor (§8.10.4.3).
⟨9⟩ $\alpha_{c,B,\infty} = \tilde{c}_B/\tilde{p}_B$ where $\tilde{c}_B$ is active substance concentration (§9.88.3).
⟨10⟩ B—Carbon dioxide; substance-concentrational solubility coefficient $= 0.230\ \mu\text{mol L}^{-1}\,\text{Pa}^{-1}$

9.102U **Units of dimension $L^{-2}M^{-1}T^2N$**
$\langle 12 \rangle$ *Coherent SI units* mole(s) per cubic metre pascal or mole second(s) squared per kilogram square metre (mol m^{-3} Pa^{-1} = mol s^2 kg^{-1} m^{-2})

9.103 Dimension $L^2MT^{-2}N^{-1}$

9.103.1 $\langle 2 \rangle$ **molar energy**

9.103.1.1 $\langle 2 \rangle$ **molar thermodynamic energy** of component B in system 1
$\langle 3 \rangle$ **molar internal energy**
$\langle 5 \rangle$ $U_{m,B}$
$\langle 6 \rangle$ thermodynamic energy of a component, U_B, divided by amount-of-substance of the component, n_B, in a system
$\langle 7 \rangle$ $U_{m,B,1} = (U_B/n_B)_1$

9.103.1.2 $\langle 2 \rangle$ **molar Gibbs energy** of reaction νB
$\langle 3 \rangle$ **affinity**
$\langle 5 \rangle$ A_B
$\langle 7 \rangle$ $A_B = -\sum\limits_{B}^{N} (\nu_B \mu_B)$ where B, ... , N are the reactants of a reaction.

9.103.1.3 $\langle 2 \rangle$ **chemical potential** of component B in system 1
$\langle 3 \rangle$ **absolute chemical potential; partial molar Gibbs energy**
$\langle 5 \rangle$ μ_B
$\langle 6 \rangle$ change in the Gibbs energy of a system, G_1, with a small addition of a component divided by the amount-of-substance of the component added, n_B, if thermodynamic temperature, pressure and the amounts-of-substance of other components are kept constant.
$\langle 7 \rangle$ $\mu_B = (dG_1/dn_{o,B})_{T,p,n(A),n(C),\ \dots\ ,\ n(N)}$
$\qquad = (dU_B/dn_{o,B})_{V,S,Q,n(A),n(C),\ \dots\ ,\ n(N)}$
$\langle 9 \rangle$ (1) Only changes in chemical potential or electrochemical potential can be measured.
$\langle 9 \rangle$ (2) Chemical potential is proportional to electric potential, E, of an ideal electrode for component B
$\qquad\qquad \Delta\mu_B = z_B F \Delta E$ where z_B is charge number (§8.1.1.3 and F is molar electrical constant (§9.105.1).

9.103.1.4 $\langle 2 \rangle$ **molar activation energy** of reaction νB
$\langle 5 \rangle$ $E_m(\nu B)$
$\langle 7 \rangle$ $E_m = R\, T^2\, d(\ln k)\, dT$
where R is the molar gas constant, k is the rate constant, and T is the thermodynamic temperature in kelvin.
$\langle 8 \rangle$ activation energy, Arrhenius activation energy; E_A, E_a
$\langle 9 \rangle$ An operationally defined kind-of-quantity expressing the dependence of the substance rate coefficient of a reaction on temperature. In enzymology, it is usually considered to be the molar energy required for a component to form an activated complex, in which a chemical bond is made or broken. In an enzyme-catalysed reaction, that reaction represents the formation of the activated enzyme–substrate complex.

9.103.1.5 ⟨2⟩ **electrochemical potential**
⟨3⟩ absolute electrochemical potential
⟨5⟩ $\tilde{\mu}_B$
⟨7⟩ $\tilde{\mu}_B = (\delta U_B/\delta n_{o,B})_{V,S,n(A),n(C),\ldots,n(N)}$
⟨9⟩ (1) Electrical charge of the system, Q, is not constant. So μ_B cannot be measured unless a non-thermodynamic convention is adopted.
⟨9⟩ (2) $\tilde{\mu}_B = \mu_B + z_B F \varphi$

9.103.1.6 ⟨2⟩ **molar energy of photons** of radiation $\delta\lambda$
⟨5⟩ $Q_{e,m}$
⟨6⟩ energy of a radiation, $Q_e(\delta\lambda)$, divided by the amount-of-substance of photons in that radiation, $n_\Phi(\delta\lambda)$.
⟨7⟩ $Q_{e,m}(\delta\lambda) = Q_e(\delta\lambda)/n_\Phi(\delta\lambda)$
⟨8⟩ photon energy; quantum energy

9.103U **Units of dimension $L^2MT^{-2}N^{-1}$**
⟨12⟩ *Coherent SI units* joule(s) per mole or kilogram square metre(s) per mole second squared (J mol^{-1} = kg m^2 mol^{-1} s^{-2})

9.104 Dimension $LM^{-2}T^2N$

9.104.1 ⟨2⟩ **molal solubility coefficient** of solute B in system 1
⟨5⟩ $\alpha_{b,B}$, $\alpha_{m,B}$
⟨7⟩ $\alpha_{b,B} = b_B/\tilde{p}_B$ where $\tilde{p}$ is active pressure (§9.41.3).
       $= \alpha_{b,B,\infty}/\gamma_B$ where γ_B is activity factor (8.10.3).
⟨9⟩ $\alpha_{b,B,\infty} = b_B/\tilde{p}_B$

9.104U **Units of dimension $LM^{-2}T^2N$**
⟨12⟩ *Coherent SI units* mole(s) per kilogram pascal or mole metre second(s) squared per kilogram squared (mol kg^{-1} Pa^{-1} = mol m s^2 kg^{-2})

9.105 Dimension TIN^{-1}

9.105.1 ⟨2⟩ **molar electrical constant**
⟨3⟩ **Faraday constant**
⟨5⟩ F
⟨6⟩ Faraday constant is the magnitude of electrical charge per mole of electrons.
⟨7⟩ $F = N_A \times e$
       $\approx 96\,485.336\,5$ C mol^{-1}; $u_c = 0.002\,1$ C mol^{-1} [ref. 15]

9.105U **Units of dimension TIN^{-1}**
⟨12⟩ *Coherent SI units* coulomb(s) per mole or ampere second(s) per mole (C mol^{-1} = A s mol^{-1})

9.106 Dimension $M^{-1}T^3I^2N^{-1}$

9.106.1 ⟨2⟩ **molar area electric conductance** of electrolyte B or ionic component B in solution 1
⟨3⟩ **molar electrical conductivity**
⟨5⟩ $\Lambda_{m,B}$, λ_B
⟨6⟩ lineic electric conductance of a solution, κ_1, divided by substance concentration of the electrolyte, c_B, in the solution
⟨7⟩ $\Lambda_{m,B} = \kappa_1/c_B$
⟨8⟩ molar conductivity; ionic conductivity

$\langle 9 \rangle$ (1) For molar electrical conductivity, the entity is taken as that of ionic charge of the component ion, $|z^{-1}|$ B. It can then be related to electric mobility, μ_{B}, by the molar electricity (or Faraday) constant, F

$$\lambda_{\mathrm{B}} = c(|z_{\mathrm{B}}| \ \mathrm{B}) \ F \ \mu_{\mathrm{B}}$$

$\langle 9 \rangle$ (2) For the usage of *molar* in terms like *molar conductivity*, see §5.4.12.

9.106U Units of dimension $\mathbf{M^{-1}T^3I^2N^{-1}}$

$\langle 12 \rangle$ *Coherent SI units* siemens square metre(s) per mole or second coulomb(s) squared per kilogram mole or ampere squared second(s) cubed per kilogram mole (S m^2 mol^{-1} = s C^2 kg^{-1} mol^{-1} = A^2 s^3 kg^{-1} mol^{-1})

9.107 Dimension $\mathbf{L^3\Theta N^{-1}}$

9.107.1 $\langle 2 \rangle$ **substance-concentrational freezing-point coefficient** of solution 1
$\langle 3 \rangle$ **concentrational freezing-point depression constant**
$\langle 5 \rangle$ $K_{\mathrm{fus},c}$, $K_{\mathrm{s}\to 1,c}$ where the subscript 'fus' is used for fusion.
$\langle 7 \rangle$ $K_{\mathrm{fus},c} = \Delta T_{\mathrm{fus}}/\hat{c}$ where $\hat{c}$ is active substance concentration of solution (§9.88.3).
$\qquad = K_{\mathrm{fus},b}/\rho_{\mathrm{A}*}$ where $\rho_{\mathrm{A}*}$ is mass concentration of pure solute (§9.14.2).

9.107U Unit of dimension $\mathbf{L^3\Theta N^{-1}}$

$\langle 12 \rangle$ *Coherent SI unit* cubic metre kelvin(s) per mole (m^3 K mol^{-1})

9.108 Dimension $\mathbf{M\Theta N^{-1}}$

9.108.1 $\langle 2 \rangle$ **molal freezing-point coefficient** of solution 1
$\langle 3 \rangle$ **molal freezing-point depression constant**
$\langle 5 \rangle$ $K_{\mathrm{fus},b}$, $K_{\mathrm{s}\to 1,b}$
$\langle 7 \rangle$ $K_{\mathrm{fus},b} = \Delta T_{\mathrm{fus}}/\hat{b}$ where $\hat{b}$ is osmolality (§9.91.2.3).
$\qquad = R \ T_{\mathrm{fus}}{}^2/\Delta H_{\mathrm{fus}}$ where H_{fus} is enthalpy of fusion (§9.42.1.8).
$\qquad = 1.855$ kg K mol^{-1}

9.108U Unit of dimension $\mathbf{M\Theta N^{-1}}$

$\langle 12 \rangle$ *Coherent SI unit* kilogram kelvin(s) per mole (kg K mol^{-1})

9.109 Dimension $\mathbf{L^2MT^{-2}\Theta^{-1}N^{-1}}$

9.109.1 $\langle 2 \rangle$ **molar kelvic enthalpy** of gaseous system 1
$\langle 3 \rangle$ **molar heat capacity**
$\langle 5 \rangle$ C_{m}
$\langle 6 \rangle$ kelvic enthalpy (or heat capacity) of a system, C_1, divided by amount-of-substance of its molecules, $\sum\limits_{\mathrm{A}}^{\mathrm{N}} n$

$\langle 7 \rangle$ $C_{\mathrm{m},1} = C_1 \Big/ \sum\limits_{\mathrm{A}}^{\mathrm{N}} n$

9.109.2 $\langle 2 \rangle$ **molar kelvic energy constant**
$\langle 3 \rangle$ **molar gas constant**
$\langle 5 \rangle$ R
$\langle 6 \rangle$ product of pressure, p, and molar volume, V_{m}, of an ideal gas divided by thermodynamic temperature, T
$\langle 7 \rangle$ $R = (p \ V_{\mathrm{m}}/T)_{\mathrm{id}}$
$\qquad \approx 8.314\,462\,1$ J K^{-1} mol^{-1}; $u_{\mathrm{c}} = 0.000\,007\,5$ J K^{-1} mol^{-1}

⟨9⟩ (1) According to the ideal gas law, the product of pressure and molar volume is proportional to thermodynamic temperature. The proportionality coefficient is the molar kelvic energy constant.

⟨9⟩ (2) The molar kelvic energy constant is related to the entitic kelvic energy constant, k (§9.73.2), by the Avogadro constant, N_A or L (§9.84.1):

$$R = N_A\, k = L\, k$$

9.109U   Units of dimension $L^2MT^{-2}\Theta^{-1}N^{-1}$

⟨12⟩ *Coherent SI units* joule(s) per kelvin mole or kilogram square metre(s) per second squared kelvin mole ($J\ K^{-1}\ mol^{-1} = kg\ m^2\ s^{-2}\ K^{-1}\ mol^{-1}$)

References

1. ICSH, IFCC and WASP, Recommendations for the use of SI in clinical laboratory measurements, *Br. J. Haematol.*, 1972, **23**, 787–788. Also *Z. Klin. Chem.*, 1973, **11**, 93; *IFCC Newsletter*, 1973, **08**(8). Es: Cinco resoluciones preparadas pos el Panel de Expertos en Unidades y Medidas (IFCC) y la Comison de Unidades y Medidas en Quimica Clinica. *Acta Bioquim. Clin. Latinoam.*, 1976, **10**(4), 362.
2. *Le Système International d'Unités (SI), The International System of Units (SI)*, 8th edn, (Bureau International des Poids et Mesures, Sèvres, France, 2006); it is the official French text followed by an English translation; commonly called the BIPM SI Brochure. Available on the Web site of the BIPM: www.bipm.org/en/si/ (acc. 2015-12).
3. R. Dybkær and K. Jørgensen. IUPAC–IFCC (International Union of Pure and Applied Chemistry, and International Federation of Clinical Chemistry, Commission on Clinical Chemistry). *Quantities and Units in Clinical Chemistry, including Recommendation 1966*. Munksgaard, København, 1967.
4. ISO/IEC 80000-9:2009. *Quantities and units — Part 9: Physical chemistry and molecular physics*. Supersedes ISO 31-8:1992.
5. ISO/IEC 80000-1:2009/ Cor 1:2011. *Quantities and units — Part 1: General*. Supersedes ISO 31-0: 1992 with the original title *Quantities and units — Part 1: General principles*.
6. ISO 2014:1976. *Writing of calendar dates in all-numeric form*.
7. ISO 8601:2004. *Data elements and interchange formats — Information interchange - Representation of dates and times*.
8. ISO/IEC 80000-3:2006. *Quantities and units — Part 3. Space and time*. Supersedes ISO 31-2:1992 with the original title *Quantities and units — Part 2. Periodic and related phenomena*.
9. B. F. Visser, *Logarithmic Quantities and Units* - O. Siggaard-Andersen, Private Press, Copenhagen, 1981, pp. 23–30.
10. ISO/IEC 80000-4:2006. *Quantities and units — Part 4. Mechanics*. Supersedes ISO 31-3:1992.
11. ISO/IEC 80000-8:2007. *Quantities and units — Part 8. Acoustics*. Supersedes ISO 31-7:1992.
12. International Commission on Radiation Units & Measurements. 1980. *Radiation quantities and units* - Report 33.
13. I. Mills, T. Cvitaš, K. Homann, N. Kallay and K. Kuchitsu, *Quantities, Units and Symbols in Physical Chemistry — The IUPAC Green Book*, RSC Publishing, Cambridge, 3rd edn, 2007.
14. ISO/IEC 80000-5:2007. *Quantities and units — Part 5. Thermodynamics*. Supersedes ISO 31-4:1992 with the original title *Quantities and units — Part 4. Heat*.
15. P. J. Mohr, B. N. Taylor and D. B. Newell, Codata recommended values of the fundamental physical constants, *Rev. Mod. Phys.*, 2012, **84**, 1527–1605. It replaces the previously recommended 2006 CODATA set. http://physics.nist.gov/cuu/Constants/codata.pdf (acc. 2015-12).
16. R. Dybkaer, IUPAC-CQUCC & IFCC-CQU (1979). List of quantities in clinical chemistry; Recommendation 1978, *Pure Appl. Chem.*, 1979, **51**(12), 2481–2502. Also *Clin. Chim. Acta*, 1979, **96**, 185F–204F.

SECTION 10

Kinds-of-property without Dimensions of the ISQ

Kinds-of-property in the International System of Quantities (ISQ) are all related to dimensions and corresponding SI units as detailed in Sections 8 and 9. These kinds-of-property have magnitude and are either differential or rational kinds-of-quantity (§4.8 and §4.11). This section concerns kinds-of-property that have no dimensions and are outside the ISQ. Definitions comply with the terminology rules of ISO 704 [1] and ISO 1087-1.[2] Note that the format developed for the presentation of measurable quantities by IUPAC-IFCC[3] (see §6.4) was generalized[4] and now applies also to dedicated kinds-of-property without dimensions of the ISQ.

10.1   Nominal Kinds-of-property

Nominal kinds-of-property are not related to magnitude, dimension or unit (§4.8). They are used in identifying and classifying objects,[5-7] defined as anything perceivable or conceivable.[1] The following examples show that many disciplines are concerned with nominal kinds-of-property.

Examples
- ☐ Pt—Cerebrospinal fluid; clarity(before spinning; visual) = turbid
- ☐ Pt(spec.)—Synovial fluid; colour(proc.) = lemon
- ☐ U—Neuroleptic drug; taxon(proc.) = chlorpromazine
- ☐ F—*Salmonella + Shigella*; taxon(proc.) = *Salmonella typhimurium*
- ☐ CsF—Virus; taxon(proc.) = Hepatitis B virus
- ☐ F—Parasite; taxon(proc.) = *Oxyuris vermicularis* eggs
- ☐ P—HLA class I antibody; taxon(HLA-A, HLA-B, HLA-C; proc.) = HLA-A
- ☐ DNA(Lkc)—FRDA gene; sequence variation(proc.) = (GAA)n expansion

NOTE 1: The abbreviation "proc." refers to the examination procedure. For nominal properties the examination procedure is an integral part of the definition of a property.
NOTE 2: The preceding examples illustrate that dedicated nominal kinds-of-property (when spatio-temporarily specified) have a value that can be expressed in words or symbols. The examined value may be expressed in a narrative fashion. For example, the morphology of lymphocytes in blood may be described in some words:
B—Lymphocytes; morphology(proc.) = sometimes large with numerous granulations

Compendium of Terminology and Nomenclature of Properties in Clinical Laboratory Sciences:
Recommendations 2016
Edited by Georges Férard, René Dybkaer and Xavier Fuentes-Arderiu
© International Union of Pure and Applied Chemistry 2017
Published by the Royal Society of Chemistry, www.rsc.org

10.2   Arbitrary Kinds-of-quantity

The modifier "arbitrary" (abbreviated "arb.") indicates that the kind-of-quantity is outside the ISQ, *i.e.*, as for kind-of-property, there is no dimension or SI measurement unit involved. In contrast to nominal kinds-of-property (§10.1), arbitrary kinds-of-quantity are related to magnitudes.

Arbitrary kinds-of-quantity may be divided into ordinal arbitrary, differential arbitrary, rational arbitrary, and other arbitrary kinds-of-quantity.

10.2.1   Ordinal Kinds-of-quantity

Ordinal kinds-of-quantity are related to magnitudes expressed by a numerical value or words denoting order of magnitude and a conventional ordinal measurement procedure (abbreviated "proc."), but have neither ISQ dimensions nor SI units. Ordinal scales are often used for "dipstick" examinations, but such measuring systems may sometimes use rational scales.

Examples

- [ ] U—Bilirubins; arb.c.(proc.;{0, 1, 2, 4}) = 2
- [ ] Trcs(B)—Aggregation, collagen-induced; arb.act.(proc.;{normal; lightly weakened; weakened; extremely weakened}) = lightly weakened
- [ ] DNA(spec.)—MTHFR gene(MIM607093.0003); arb.entitic num.(proc.;{0, 1, 2}) = 1
- [ ] DNA(spec.)—CFTR gene(MIM602421.0005); arb.entitic num.(proc.;{0, 1, 2}) = 1
- [ ] DNA(spec.)—HLA-B gene(B27); arb.entitic num.(proc.;{0, 1}) = 1
- [ ] P—Cat epithelium antibody(IgE); arb.c.(proc.;NCCLS/e1{neg, pos}) = negative

NOTE 1: The scale is defined in the measurement procedure.

NOTE 2: The plural "bilirubins" indicates the sum of the neutral and ionic forms of bilirubin.

10.2.2   Differential Arbitrary Kinds-of-quantity

Differential arbitrary kinds-of-quantity have magnitudes and can be subtracted from, but cannot be divided by another quantity of the same kind-of-quantity.

10.2.3   Rational Arbitrary Kinds-of-quantity with WHO International Units

Some rational arbitrary kinds-of-quantity, usually with biological components, have international units (here symbolized IU), defined by WHO. Each international unit is defined by a certified reference material (CRMs, see §6.11.4), which should be indicated in the specification of the kind-of-property. See examples below. No dimension is defined.

The kind-of-quantity is identified by the preceding "arbitrary" and parenthetic "procedure".

Examples

- [ ] P—Protein S; arb.subst.c.(IS 93/590; proc.) = 1.1×10^3 IU L^{-1}
- [ ] P—Insulin; arb.subst.c.(IRP 66/304; proc.) = 19×10^{-3} IU L^{-1}
- [ ] P—Olive antibody(IgE); arb.subst.c.(IRP 75/502; NCCLS/t9; proc.) = 2.2×10^3 IU L^{-1}
- [ ] P—DNA(double stranded) antibody(IgG); arb.subst.c.(IS WHO/80; proc.) = 9×10^3 IU L^{-1}

10.2.4   Other Arbitrary Kinds-of-quantity

Some arbitrary kinds-of-quantity, defined by their measurement procedures, have units defined there. The IUPAC-IFCC format indicates this situation by specifying "procedure defined unit" (abbreviated "p.d.u."); both term and symbol are language dependent.

Examples

☐ B—Sedimentation reaction; arb.length(proc.) = 8 mm (p.d.u.)
☐ CsF—Adenovirus antibody(IgG); arb.subst.c.(proc.) = ? (p.d.u.)
☐ P—Plasminogen activator inhibitor 2; arb.subst.c.(imm.; proc.) = ? (p.d.u.)

NOTE 1: Some of the kinds-of-quantity in §10.2.3 and §10.3.1 can be redefined to have ISQ dimensional kinds-of-quantity if their magnitudes are proportional to the number of molecules of the component. It is also applicable to so-termed "serological titre".

NOTE 2: If no reference for the unit is given, the unit is undefined. At the place for the unit is stated "procedure defined unit", abbreviated p.d.u. Note that the term "p.d.u." designates a unit of unknown magnitude.

10.3 Arbitrary Biological Activities

10.3.1 ⟨2⟩ **vitamin activity** of component B in foodstuff 1 for organism 2
⟨3⟩ **vitamin equivalent**
⟨6⟩ vitamin activity of a component in a foodstuff expressed as amount of a reference component in a comparable foodstuff giving equal activity by a defined measurement procedure
⟨9⟩ The International Union of Nutritional Sciences defines niacin activity in terms of mass of nicotinic acid giving the same biological activity. For example, niacin activity is then termed the **generic activity** of nicotinic acid and nicotinamide. The concept can alternatively be expressed as the corresponding amount-of-substance of nicotinic acid. For human nutrition, factors have been published relating the activities of different components with a generic vitamin activity.

10.3.2 ⟨2⟩ **toxicity activity** of component B in environment 1 for organism 2
⟨6⟩ amount-of-substance of a component or a group of components in an environment expressed as the amount-of-substance of a reference component in a comparable environment that has the same toxic effect, as measured by a defined procedure
⟨9⟩ The result of such measurement procedures can be expressed as the corresponding amount-of-substance (or substance concentration or substance rate) of a reference component.

References

1. ISO 704:2000. *Terminology work — Principles and methods.*
2. ISO 1087-1:2000. *Terminology work — Vocabulary — Part 1: Theory and application.*
3. R. Dybkær and K. Jørgensen, IUPAC–IFCC (International Union of Pure and Applied Chemistry, and International Federation of Clinical Chemistry, Commission on Clinical Chemistry). *Quantities and Units in Clinical Chemistry, including Recommendation 1966.* Munksgaard, København, 1967.
4. ISO 15189:2007. *Medical laboratories — Particular requirements for quality and competence.* Replaced by ISO 15189:2012.
5. BIPM; IEC; IFCC; ILAC; ISO; IUPAC; IUPAP; OIML, 2012. *International vocabulary of metrology – Basic and general concepts and associated terms VIM*, 3rd edn. JCGM 200: 2012. This 3rd edition is also published as ISO Guide 99 by ISO (ISO/IEC Guide 99-12: 2007). Replaces the 2nd edition, 1993. Available on the Web site of the BIPM: http://www.bipm.org, (acc. 2015-12).
6. R. Dybkaer, *An Ontology on property for physical, chemical, and biological systems*, 2009, http://ontology.iupac.org, (acc. 2015-12).
7. R. Dybkaer, IUPAC-CQUCC & IFCC-CQU (1979). List of quantities in clinical chemistry; Recommendation 1978, *Pure Appl. Chem.*, 1979, **51**(12), 2481–2502. Also *Clin. Chim. Acta*, 1979, **96**(1–2), 185F–204F.

Index of Tables and Figures

Tables

Table 4.1 Classification of types of property with their algebraic delimiting characteristics.

Table 4.2 Other basic concepts involved in the production of an examination result or a measurement result.

Table 5.1 Base kinds-of-quantity, base units, and their dimensional symbol (§5.2) in the International System of Units (SI). The status of number of entities is also that of a base kind-of-quantity [ref. 1, concept 1.16, Note 4].

Table 5.2 Derived units of the International System of Units (SI) with special terms or symbols. The sequence of the list is as in Sections 8 and 9, essentially by order of dimension (Table 5.1) and increasing powers of those dimensions, first positive and then negative. For electrical and luminous kinds-of-quantity, systematic nomenclature of the kinds-of-quantity is based on electrical charge (unit $C = A\,s$) and amount-of-light (unit $lm\,s = cd\,sr\,s$). The term and symbol for the katal have been recognized by IUPAC and IFCC,[6] IUB,[7] WHO,[8] and finally by CGPM (1999, Resolution 12). Note that the symbols t, ϑ and Φ have several meanings in Table 5.1, those of Φ being distinguished here by subscripts. N (Table 5.1) is also to be distinguished from N (this table).

Table 5.3 SI prefixes denoting decimal factors, 10^n. Da, Danish; Es, Spanish; Gr, Greek; It, Italian; La, Latin; No Norwegian. m, exponent of 10^3.

Table 5.4 Non-SI units accepted for use together with the International System of Units (SI).[10] For experimentally obtained values of dalton and electronvolt, standard measurement uncertainty (u) is stated. The unified atomic mass unit (u) is recognized by CGPM but the term dalton and symbol Da are preferred by IUPAC-IUB;[11] results can however alternatively be expressed as molar mass (§9.8U, ⟨19⟩ (3)). Other units are recognized by CGPM[10] and mentioned in the SI brochure [ref. 10, Table 6]; those marked here with an asterisk (*) are mentioned in the SI brochure [ref. 10, Table 6] as in "common everyday use" and are in national legislation of most countries.

Table 6.1 Abbreviations for systems in the human body. They were developed for the English language[10] and by the Danish data bank of dedicated kinds-of-property in clinical laboratory sciences.[11] The distinction between singular and plural (by an s) may be used to indicate whether the object of the study is a single entity or a collection of entities.

Table 6.2 English-language abbreviations for kinds-of-property used in clinical laboratory sciences.

Compendium of Terminology and Nomenclature of Properties in Clinical Laboratory Sciences:
Recommendations 2016
Edited by Georges Férard, René Dybkaer and Xavier Fuentes-Arderiu
© International Union of Pure and Applied Chemistry 2017
Published by the Royal Society of Chemistry, www.rsc.org

Table 7.1 Terms and symbols of compositional kinds-of-quantity, derived from two extensive kinds-of-quantity Q and Q': number, volume, mass, amount-of-substance, catalytic activity and absorbed dose, $q_B = Q_B/Q'_1$. For the meaning of modifier symbols, see §4.3.1.

Table 7.2 Terms and symbols of material kinds-of-quantity (fundamental constants, material constants or material coefficients) derived from two extensive kinds-of-quantity Q and Q': number, volume, mass and amount-of-substance, $q_B = Q_B/Q'_B$ or $q_1 = Q_1/Q'_1$. They can be used as coefficients of proportionality to convert compositional kinds-of-quantity listed in Table 7.1. For the meaning of modifiers, see §4.3.14.

Table 7.3 Terms of regions of the spectrum of electromagnetic radiation in terms of wavelength in vacuum, λ_0, wavenumber in vacuum, $\tilde{\nu}$, frequency, ν, entitic energy, Q_e/N_Φ, and molar energy, Q_e/n_Φ; h, Planck constant; N_A, Avogadro constant; n, amount-of-substance; N, number of entities; subscript Φ, photons; Q_e, energy of the radiation. The broken lines indicate that the boundaries are arbitrary and are differently defined by different authorities, including ir., infrared; uv, ultraviolet.

Tables of units for each kind-of-quantity in Sections 8 and 9 are not listed here.

Figures

Figure 2.1 Generic concept diagram around 'clinical biology' = 'clinical laboratory sciences'. A short generic relation line to three bullets indicates that one or more specific concepts are possible.

Figure 4.1 Two generic concept systems as tree diagrams.

Figure 6.1 Elements of a clinical laboratory report.

Subject Index